D0394748

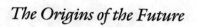
The Origins of the Future

Gribbin, John R.
The origins of the
future : ten questions f
c2006.
33305211979939
la   12/27/06

The Origins of the Future

Ten Questions for the Next Ten Years

John Gribbin

Yale University Press   New Haven and London

Copyright © 2006 by John and Mary Gribbin.
All rights reserved.
This book may not be reproduced, in whole or in part, including illustrations, in any form
(beyond that copying permitted by Sections 107 and 108 of the U.S. Copyright Law and except by reviewers for the public press), without written permission from the publishers.

Set in Galliard type by Keystone Typesetting, Inc.
Printed in the United States of America by R. R. Donnelley, Harrisonburg, Virginia.

Library of Congress Cataloging-in-Publication Data
Gribbin, John R.
The origins of the future : ten questions for the next ten years / John Gribbin. — 1st ed.
p. cm.
Includes bibliographical references and index.
ISBN-13: 978-0-300-11998-5 (alk. paper)
ISBN-10: 0-300-11998-4 (alk. paper)
1. Cosmology — Miscellanea. 2. Evolution — Miscellanea. 3. Life — Origin. I. Title.
QB981.G765 2006
523.1 — dc22   2006011062

A catalogue record for this book is available from the British Library.

The paper in this book meets the guidelines for permanence and durability of the Commit-
tee on Production Guidelines for Book Longevity of the Council on Library Resources.

10 9 8 7 6 5 4 3 2 1

In memory of Fred Hoyle (1915–2001)

When a thing was new, people said, "It is not true." Later, when the truth became obvious, people said, "Anyway, it is not important." And when it's importance could not be denied, people said, "Anyway, it is not new."
— William James (1842–1920)

It is not *what* the man of science believes that distinguishes him, but *how* and *why* he believes it. His beliefs are tentative, not dogmatic; they are based on evidence, not on authority or intuition.
— Bertrand Russell (1872–1970)

Contents

Preface

Origins

How did the Universe begin? Where do the material particles we are made of come from? Where do galaxies come from? How do stars and planets form? How did life begin?

Most of the big questions in science today concern origins. We have only provisional answers to those questions (some more provisional than others), but they are answers which are all likely to be improved dramatically by scientific progress over the next ten years. The provisional answers we already have are vastly better than no answers at all, and the story of how we arrived at those answers is one worth telling in its own right, as well as a good way to set the scene for the stories likely to make headlines over the next decade.

When I wrote my history of science since the sixteenth century (*The Scientists*, 2002), I ended the story in the late twentieth century, stopping at slightly different times for different areas of science. It was always my intention to come back to the story and bring it up to date in some way, although clearly it would have been impossible to deal with the whole breadth of modern science in the same depth in a reasonable space. I decided to concentrate on the physical sciences, where my own experience of research is relevant, and originally intended to use the biographical approach of my earlier book, hoping to give insight into the way science works today by focusing on individual contributions. But they do science differently these days. The more I visited scientists active today the more I realized just how much science has changed, even in my own lifetime. I didn't expect to encounter giants like Newton or Darwin, but I did expect to encounter people like the scientific heroes of my youth: Fred Hoyle, Francis Crick, Richard Feynman. These were people with an extraordinary breadth of knowledge, as well as deep insight into particular problems and dramatic personal backgrounds. But even the men and women who are pushing the physical sciences forward at the cutting edge of research today are more, for want of a better word, ordinary. There is an established route to becoming "a scientist," through school and undergraduate studies to a Ph.D. and beyond, and to give biographical details of the lives of most scientists would be no more interesting than to give biographical details of the lives of most accountants. There are still great men — and women — involved in science, but now they almost all have to work in large teams, and they come from similar educational backgrounds.

But although something has been lost, something has been gained. Individual scientists in these fields generally focus on relatively small problems while working in rather large teams, so that it is often difficult to identify their specific contributions or to claim that the project would not have succeeded as well if someone else had been doing their job. Yet the whole has become greater than the sum of its parts, so that a compelling, consistent, and very nearly complete picture of how the physical world works, and how the Universe we see around us got to be the way it is, has emerged. One of the most fascinating features of this story is that the mystery of the origin of life now falls squarely within the province of physical science (and is considerably less mysterious than it used to be).

It is only by standing back from the individual contributions and looking at the big picture that the scale of this achievement becomes apparent. There *is* a fascinating scientific biography to be written, but it is the biography of the Universe itself, not of the people probing its remaining mysteries today. So, somewhat to my own surprise, that is the story that began to emerge as I put the information gleaned from my travels in order. Since the life of the Universe is far from over, I have chosen to present the story so far in terms of beginnings, from the beginning of the Universe as we know it some 14 billion (10^9) years ago to the beginning of life on Earth roughly ten billion years later. I was also unable to resist one peek into the future, to look at the likely fate of the Earth, which is of considerable interest to residents of our planet, and of the Universe at large.

Some of the outlines of this story have been sketched before (in my own *Genesis,* 1981, and in works by others), but what is different about science in the twenty-first century is that the outlines have been filled in with precision (and sometimes changed dramatically in the process) so that key numbers that define the state of the Universe are known to an accuracy of a few percent, or even to fractions of 1 percent. At the same time, different parts of the picture are found to match up with astonishing precision, so that, for example, the measured properties of the neutron (a component of the nucleus of an atom) in the laboratory are intimately related to events that occurred in the Big Bang, as well as to the amount of helium found in stars today. In turn, the amount of helium in stars affects the production of chemical elements that have ended up in your body and is relevant to the story of the origin of life.

This book deals largely with things we *think* we know about the Universe, and our place in it, rather than the near-certainties of my earlier book. Precisely because I am dealing with research at the cutting edge, some of the things I will tell you about in this book will turn out to be wrong. It would be extremely disappointing if they didn't, because that would mean nature no longer had any capacity to surprise us. But whatever the fate of these specific theories and hypotheses, it now seems extremely unlikely that the broad picture of how things got to be the way they are will change. The picture that unfolds over the next ten years or so is likely to be as spectacular for its coherence and completeness as for the significance of any particular piece of the puzzle, and my aim in this book is to provide you with the background knowledge that will help you appreciate the unfolding vision of the cosmos. Before we can come to grips with this new understanding of the Universe and our place in it, however, we need to take stock of the things physicists are sure they know about the way the Universe works, and to appreciate the difference between what we think we *know*, and what we *think* we know.

Acknowledgments

Most of the research for my recent books has involved delving in dusty archives and reading secondhand (at best) reports about the lives and work of dead people. Intriguing though this has been, it has made a pleasant change, in working on the present book, to be talking to living people about their own work. But because my aim here is to give an overview of what is going on in the physical sciences today, I have seldom referred to individuals or individual pieces of research by name in the main text. If there is one thing I have learned from writing my more historical books, it is that science is a communal activity. The "we" that appears in the text refers to the whole pool of scientists, past and present, who have contributed to our understanding of the physical Universe. But I could not have written this book without discussions and correspondence with many individual researchers, both specifically for this book and on other occasions over the years, and I would like to thank Kevork Abazajian, John Bahcall, John Barrow,

Frank Close, Ed Copeland, Pier-Stefano Corasiniti, John Faulkner, Ignacio Ferreras, Simon Goodwin, Ann Green, Ben Gribbin, Alan Guth, Martin Hendry, Mark Hindmarsh, Gilbert Holder, Isobel Hook, Jim Hough, Steve King, Chris Ladroue, Ofer Lahav, Andrew Liddle, Andrei Linde, Jim Lovelock, Gabriella De Lucia, Mike MacIntyre, Ilia Musco, Jayant Narlikar, Martin Rees, Leszek Rozkowski, B. Sathyaprakash, Richard Savage, Peter Schröder, Uros Seljak, Lee Smolin, Adam Stanford, Paul Steinhardt, Christine Sutton, Peter Thomas, Kip Thorne, Ed Tryon, Neil Turok, and Ian Waddington for their generous willingness to share their ideas with me. Going back even farther, I owe an unrepayable debt to several scientists who are no longer with us; in the order in which they exerted their influence on me, these were Bill McCrea, Fred Hoyle, Willy Fowler, Roger Tayler, and John Maynard-Smith.

Thanks also to Christine and David Glasson, who were responsible for making me take occasional breaks from the work; to the Alfred C. Munger Foundation, which made a generous contribution to my travel and other research expenses; and to the University of Sussex, which provided a base from which to work.

As always, though, the greatest behind the scenes contribution came from my ever-present but not always visible collaborator, Mary Gribbin.

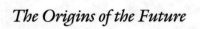

The Origins of the Future

How Do We Know the Things We Think We Know?

What do scientists mean when they say that they "know" what goes on inside an atom, say, or what happened in the first three minutes of the life of the Universe? They mean that they have what they call a model of the atom, or the early Universe, or whatever it is they are interested in, and that this model matches the results of their experiments or their observations of the world. Such a scientific model is not a physical representation of the real thing, the way a model aircraft represents a full-scale aircraft; it is a mental image which is described by a set of mathematical equations. The atoms and molecules that make up the air that we breathe, for example, can be described in terms of a model in which we imagine each particle to be a perfectly elastic little sphere (a tiny billiard ball), with all the little spheres bouncing off one another and the walls of their container.

That is the mental image, but this is only half the model; what makes it a *scientific* model is that the way the spheres move and bounce off one another is described by a set of physical laws, written in terms of mathematical equations. In this case, these are essentially the laws of motion discovered by Isaac Newton more than three hundred years ago. Using those mathematical laws, it is possible to predict, for example, what will happen to the pressure exerted by a gas if it is squashed into half its initial volume. If you do the experiment, the result you get (in this case, the pressure will double) matches the prediction of the model, which makes it a good model.

Of course, we should not be surprised that the standard model of a gas which describes it in terms of little balls bouncing off one another in accordance with Newton's laws makes this particular correct prediction, because the experiments were done first and the model was designed, or constructed, to match the results of the experiments. The next stage in the scientific process is to use the model you have developed from measurements carried out in one set of experiments to make predictions (precise, mathematical predictions) about what will happen to the same system when you do different experiments. If the model makes the "right" predictions under the new circumstances, this shows that it is a good model. Even if it fails to make the right predictions it may not have to be completely discarded, because it still tells us something useful about the earlier experiments; but at best it has restricted applicability.

In fact, all scientific models have restricted applicability. None of them is *the* truth. The model of an atom as a perfectly elastic little sphere works fine for calculating changes in pressure of a gas under different circumstances, but if you want to describe the way an atom emits or absorbs light, you need a model of the atom in which it has at least two components, a tiny central nucleus (which can itself, for some purposes, be regarded as a perfectly elastic little sphere) surrounded by a cloud of electrons. Scientific models are representations of reality, not the reality itself, and no matter how well they work or how accurate their predictions are under the appropriate circumstances, they should always be regarded as approximations and aids to the imagination, rather than as the ultimate truth. When scientists tell you that, say, the nucleus of an atom is made up of particles called protons and neutrons, what they should really say is that the nucleus of an atom behaves, under certain circumstances, *as if* it were made up of protons and neutrons. The better scientists take the "as if" as read but appreciate that their models are, indeed, only models; lesser scientists often forget this crucial distinction.

Lesser scientists, and many nonscientists, also have another misconception. They often think that the role of scientists today is to carry out experiments which will prove the accuracy of their models to better and better precision — to more and more decimal places. Not at all! The reason for carrying out experiments which probe previously untested predictions of the models is to find out where the models break down. It is the cherished hope of the best physicists to find flaws in their models, because those flaws — things that the models cannot predict accurately or cannot explain in detail — will highlight the places where we need a new understanding and better models.

The archetypal example of this is gravity. Isaac Newton's law of gravity was regarded as the most profound piece of physics for more than two hundred years, from the 1680s to the beginning of the twentieth century. But there were a few, seemingly tiny, things that the Newtonian model could not explain (or predict), involving the orbit of the planet Mercury and the way light is bent when it goes past the Sun. Albert Einstein's model of gravity, based on his general theory of relativity,* explains everything that Newton's model explains, but it *also* explains these subtle details of planetary orbits and light bending. In that sense, it is a better model than the older model, Newton's, and it makes correct predictions (in particular, about the Universe at large) that the older model does not. But Newton's model is still all you need if you are calculating the flight of a space probe from the Earth to the Moon. You *could* do the same calculation using general relativity, but it would be more tedious and it would give you the same answer, so why bother?

Most of this book will be about things we *think* we know — models which look good as far as they have been tested, but which involve emerging science, for which many more tests remain to be carried out. It is certain that some of these models will require modification in light of further experiments and observations of the Universe; it is possible that some of them may have to be completely discarded and replaced by new ways of looking at things. But to set the scene for our description of where science is going in the twenty-first century, we need to start from the things we think we *know* — models, essentially developed in the twentieth century, that have been so successful at matching the results of experiment and observation that scientists have almost as much faith in them as they have in the billiard ball model of gases or (within its known limitations) in Newton's model of gravity. These are models which are, like Newton's model, very nearly perfect in their description of the physical Universe within the specific areas for which they are known to apply; just as important,

---

*The term *theory* is often used to describe what I call models. In general, I prefer *model* because for non-scientists it carries with it less misleading baggage than *theory;* but in some cases, notably Einstein's theory, the word is so much part of the name that it cannot be avoided. Everything I have said about scientific models, however, applies also to scientific theories.

again as with Newton's model, we know where the limits of the applicability of those models are.

Physicists like to refer to those highly successful descriptions of the world (or rather, of specific features of the world) as "standard" models. The billiard ball model of gases (which is also known as the kinetic theory, since it deals with particles in motion) is a standard model. But when physicists talk of *the* standard model, they are referring to one of the great triumphs of twentieth-century science, the model which describes the behavior of particles and forces on subatomic scales. And it began, essentially, in the second decade of the twentieth century, when the Danish physicist Niels Bohr came up with a new model of the atom. I have described the historical development of quantum physics in my book *In Search of Schrödinger's Cat* (1984) and do not intend to go into details here; but since the standard model of particle physics is entirely based on quantum physics, a brief recap is in order. At first sight, some of this may seem familiar to some readers; but bear with me, because I hope you will find that my take on this familiar story is not quite the same as what you think you know.

The first step toward this new understanding of physics came from Max Planck in Germany at the beginning of the twentieth century. Planck discovered that the only way observations of the way light is radiated by hot objects could be explained was if the light were emitted in little lumps, packets called quanta. At that time, scientists thought of light as a form of wave, an electromagnetic vibration, because observations of the behavior of light in many experiments matched the predictions of the wave model. At first, neither Planck himself nor his contemporaries thought that light *existed* in the form of little lumps, only that the properties of matter — that is, of atoms — meant that it could only be emitted (or absorbed) in certain amounts. You could make an analogy with a dripping tap. The fact that water drips from the tap in the form of tiny "lumps" doesn't mean that the water in the tank feeding the tap exists only in the form of separate drops. Albert Einstein, in 1905, was the first person in modern times to take seriously the idea that light actually existed in the form of little lumps, particles of light that came to be known as photons, and for roughly the next ten years he was essentially in a minority of one.* But it turns out that the behavior of light in some experiments actually matches the predictions of the particle model. So the particle model must be a good one as well. No experiment shows light behaving like a wave and like a particle at the same time; but light can match the predictions of either model depending on the nature of the experiment.

It's worth getting this clear, because it is such a fine example of the limitations of models. Nobody should ever have said (or thought) that light *is* a wave, or *is* a particle. All we can say is that under appropriate circumstances light behaves *as if* it were a wave or *as if* it were a particle — just as under certain circumstances an atom behaves *as if* it were a little hard ball, while under other circumstances it behaves *as if* it were a tiny nucleus surrounded by a cloud of electrons. There is no paradox or conflict here. The limitations lie in our models and our human imagination, because we are trying to describe something that is, in its entirety, unlike anything we have experienced with our own senses. The confusion we feel when we are trying to imagine how light can be both a wave and a particle is part of what the American physicist Richard Feynman called "a reflection of an uncontrolled but vain desire to see it in terms of something familiar." Light is

---

*Newton had offered a respectable particle model of light, but it had been superseded by the wave model.

actually a quantum phenomenon which can be described very effectively in terms of mathematical equations but for which no single mental image from everyday life will do to give us an idea of what it is like. The whole quantum world is like that, and Niels Bohr's first great contribution to physics was to incorporate the mathematics of quantum physics into a model of the atom without worrying too much about whether that model made sense in everyday terms.

By the beginning of the twentieth century, scientists knew that everything on Earth is made of atoms and that there is one kind of atom for each of the chemical elements — atoms of oxygen, atoms of gold, atoms of hydrogen, and so on. They also knew that atoms were not indivisible, as had once been thought, but that pieces, called electrons, could be chipped off them under the right circumstances. At that time, the favored model of the electron described it as a tiny particle, and experiments had shown that electrons did, indeed, behave as if they were tiny particles. The puzzle Bohr solved was the way in which light is radiated (or absorbed) by different kinds of individual atoms; furthermore, his model explained the puzzle at a more detailed level than had Planck's study of the light emitted by glowing objects made up of many different atoms. If a pure element (such as sodium) is heated in a flame, it radiates at precise wavelengths, or colors, which produce a pattern of lines in the spectrum of visible light, which spans all the colors of the rainbow (indeed, a rainbow *is* a spectrum). In the case of sodium, the lines are in the orange-yellow part of the rainbow; but each element (which means each kind of atom) produces its own distinctive pattern of lines, as unique as a fingerprint and resembling a barcode. The blend of colors seen in a rainbow is the result of many different kinds of atom all radiating at different wavelengths. This blend usually ends up giving the appearance of white light, but the colors are separated by the way light is bent within raindrops, or, as Isaac Newton's investigations showed, by a triangular prism of glass.

Now, light is a form of energy, and the energy in light emitted by atoms must come from within the atoms (it is the most basic law of physics that energy cannot be created out of nothing, although even that rule has its limitations, as we shall see). Bohr realized that the energy comes from rearranging the electrons in the outer part of an atom. (Light is, therefore, literally atomic energy, and what has become known, through a quirk of history, as atomic energy should really be called nuclear energy.) Electrons carry negative charge, while the nucleus of an atom carries positive charge, so electrons are attracted to the nucleus in a way similar to the way an object on Earth is attracted by gravity to the Earth itself. If you carry a weight upstairs, you have to do work (put in energy) to move the weight farther away from the center of the Earth. If you drop the weight out of an upstairs window, this energy is released, converted first into the energy of motion of the falling weight and then, when it hits the ground, into heat, warming the ground slightly as it makes the atoms and molecules at the impact site jiggle about. Bohr suggested that in the same sort of way if an electron in the outer part of an atom moved closer to the nucleus, energy would be given up (in this case, as light). If an electron closer to the nucleus absorbed energy (perhaps from light, or because the atom was being heated), it would jump farther out. But why should the energy be emitted and absorbed only at precise wavelengths, corresponding to precise amounts of energy?

In the model developed by Bohr, an electron is envisaged as moving around the nucleus in a way reminiscent of the way planets orbit around the Sun. But whereas a planet can orbit, in principle, at any distance from the Sun, only certain electron orbits, according to Bohr, were "allowed"—rather as if a planet could occupy the orbit of the Earth or the orbit of Mars, but not anywhere in between. Then, he suggested, an electron could jump from one orbit to another (as if Mars jumped into the Earth's orbit) and emit a precise amount of energy (corresponding to a precise wavelength of light) in the process. But it could not jump to an in-between orbit, and emit an in-between amount of energy because there were no in-between orbits. Of course, he supplied mathematics to back this up, based on studies of spectra, and the physics was developed by further experiments and observations. What matters, however, is that Bohr found a model which could predict where the lines in atomic spectra should be, even though the idea of what became known as "quantized" orbits made no sense in terms of our everyday experience. Equally baffling, according to this model changes occur as if an electron disappears from one orbit and instantly appears in another orbit without ever crossing the space in between. Although it took a long time for scientists to grasp the point, Bohr had made it clear that a model doesn't have to make sense in order to be a good model; all it has to do is make predictions (based on sound mathematics and observed physics) that match the outcome of experiments.

Bohr's model of the atom is often regarded as rather quaint and old-fashioned today. The physicists' image of the electron has changed a lot since his day, not least since the discovery in the 1920s that under some experimental circumstances an electron behaves as if it were a wave. Just like light (and, indeed, like every other entity in the quantum world) there is a "wave-particle duality" about the electron. We cannot say that it *is* a wave, or that it *is* a particle, only that sometimes (in a predictable way, not on a whim) it behaves as if it were a wave and sometimes it behaves as if it were a particle. This discovery led to the idea that all the electrons in an atom occupy a fuzzy, diffuse cloud around the nucleus; changes in the energy of the cloud occur in more subtle ways than by a tiny particle hopping from one orbit to another. This is a more sophisticated model, which works very well when we want to explain how atoms join together to make molecules and thereby underpins our entire modern understanding of chemistry. But just as Newton's model is all you need to know about gravity if you are calculating the trajectory of a space probe going to the Moon, so Bohr's model still works if all you want to do is explain the lines you see in the spectra of hot materials, such as sodium (or, indeed, the Sun). Old models seldom die; they just have restricted usefulness.

That's all we have to say about electrons for now, because in the standard model an electron is regarded as one of the basic building blocks of matter — a truly fundamental entity which is not made up of smaller things. But the same is not true of the nucleus. And as well as "explaining" — that is, providing us with a working model to describe — what the nucleus is, the standard model also offers insight into the forces that operate between the kinds of fundamental entities we are used to thinking of as particles.

It's hard to avoid using the term *particle* at least some of the time when talking about fundamental entities such as electrons, and we shall not always qualify it. But it is important to remember that our use of the word doesn't mean that these entities should be thought of solely as little hard spheres or concentrations of mass and energy at a point. They do behave like that in some experiments, but they don't in others. The term *wavicle* has sometimes been used to try to convey a flavor of the dual wave-particle nature of quantum objects, but we are not convinced that this works. On the other hand, physicists do have a perfectly good alternative to the word *force,* which is just as well, since quantum "forces" are as strange, in everyday terms, as quantum "particles."

We are all familiar with two forces of nature — gravity and electromagnetism. We feel the Earth pulling us down, and we have seen a magnet picking up a metal object, or charged up a plastic comb by running it through our hair and then used the static electricity generated in it to pick up tiny pieces of paper. But as those examples show, forces always operate between two (or more) objects — the Earth pulls us down, the magnet picks up a nail. There is an interaction between the objects involved, and that gives physicists their preferred term, *interaction,* to describe what is going on. It might seem, from the examples we have given, that there are three interactions experienced in everyday life, because magnetism and electricity have superficially different properties. But in the nineteenth century the Scottish physicist James Clerk Maxwell, building on the work of Londoner Michael Faraday, discovered a set of equations that describe both electricity and magnetism within the framework of a single model. They are actually different aspects of the same interaction, like the two faces of a coin.

There are, though, several genuine and important differences between the gravitational interaction and the electromagnetic interaction. Gravity is very, very much weaker than electromagnetism. It takes the pull of the entire Earth to hold a steel pin down on the ground, for example, but a child's toy magnet can easily overcome this pull and lift the pin upward. Because electrons and atomic nuclei carry electric charge, and the strength of the gravitational pull of one single atom on another is so tiny it can be ignored, all the significant interactions between atoms are electromagnetic. So electromagnetic forces hold your body together and make your muscles work. If you pick an apple up off the table, electromagnetic interactions in your muscles are overpowering the gravitational interaction between the apple and the Earth. In a real sense, you are more powerful than a planet, thanks to electromagnetic interactions.

But although gravity is weak, it has a very long range. The interaction between the Sun and the planets holds the planets in their orbits, and in a similar way the Sun itself is part of a system of several hundred billion stars forming a disk-shaped galaxy roughly a hundred thousand light years across, rotating around its center and held together by gravity. In principle, electromagnetic interactions are just as long-range. But another difference between them and gravity is that they come in different varieties, which cancel each other out. In an atom, the positive charge of the nucleus is cancelled by the negative charge of the electrons, so from any great distance — great compared with the size of the atom — it seems to be electrically neutral, with no overall charge. North magnetic poles, similarly, are always accompanied by south magnetic poles, and although the magnetic fields of objects like the Sun and the Earth do extend into space to some extent, on cosmic scales there is no overall magnetic influence pulling things together or pushing them apart.

That's another thing about electromagnetism that distinguishes it from gravity. Gravity always attracts. But although opposite electric charges and opposite magnetic poles attract, similar charges and similar poles repel one another, something we all discovered as children when we tried to squash the two north poles of two different magnets together. So even before they tried to investigate the quantum realm, physicists knew that interactions (forces) could have long or short range, that they could be associated with different kinds of "charge," and that they could either attract or repel. More subtly, we see that not all interactions affect everything in the same way. Gravity seems to be universal and does affect everything. But electric and magnetic influences only affect certain kinds of objects. All of this came in useful when the physicists began to probe inside the nucleus.

The way they probed within the nucleus was by firing beams of —

for want of a better word—particles at nuclei and subnuclear particles, and measuring the way they bounced off. The more energy there is in the incoming particles, the finer the details that can be resolved in the "target." At first, early in the twentieth century, this was done using particles produced naturally by radioactive processes. As technology developed, this technique was refined to take particles such as electrons and accelerate them to very high energies, using magnetic fields in machines called (logically enough) particle accelerators. This in turn led to the development of huge accelerators like those at CERN, near Geneva, where cutting-edge research into the nature of matter and the interactions ("forces of nature") continues today.

After the discovery of the nucleus itself, in experiments carried out in Cambridge at the beginning of the 1910s, the next step came with the discovery in the 1920s that the nucleus responds to this probing as if it were a ball made of two kinds of particles, protons and neutrons, jammed together like a closely packed cluster of grapes. The nucleus of the simplest atom, hydrogen, actually consists of a single proton, but all other nuclei contain neutrons as well as protons—the most common form of uranium, for example, contains 92 protons and 146 neutrons. Each proton has an amount of positive charge equal in size to the amount of negative charge on a single electron, so in a neutral atom there are exactly the same number of protons as electrons. Each neutron, as the name implies, is electrically neutral. The obvious question was, Why doesn't the repulsion of the interaction between all the positively charged protons blow the nucleus apart? The obvious answer, later borne out by experiments, was that there must be a previously unsuspected attractive interaction that overwhelms the electric repulsion and holds the nucleus together. Because this interaction is stronger than the electromagnetic interaction, it became known as the strong interaction (or the strong force). And because no trace of its influence can be detected far away from the nucleus, it was clear that it must have a short range, extending its influence only over the diameter of a large nucleus. This is why there are no nuclei much bigger than that of uranium. If you want a pictorial model, imagine trying to jam more than about 240 protons and neutrons together: the protons on opposite sides of the ball will still be repelled strongly from one another by the electromagnetic interaction, but they will be too far apart to feel the attraction of the strong interaction.

The energies required to probe inside protons and neutrons (collectively known as nucleons) are so big that it took decades — from the 1930s to the 1960s — to come up with a reliable model of what goes on inside these particles. The picture that emerged was consistent with a model in which each of the nucleons is composed of three truly fundamental entities (as fundamental as electrons), which were called quarks. Experiments probing protons and neutrons supported the predictions of a model in which there were two kinds of quark, dubbed "up" and "down"; the proton is regarded as composed of two up quarks and one down quark, while the neutron is composed of two down quarks and one up quark. With a charge of $\frac{1}{3}$ of the charge on an electron allotted to each down quark, and $\frac{2}{3}$ of the charge on a proton allotted to each up quark, the numbers add up to explain the observed charges of protons and neutrons.

But why have individual quarks, or any particles with "fractional" charge, never been detected? The model explains this (and experiments back the explanation up) by proposing that pairs or triplets of quarks are "confined" within composite particles such as protons and neutrons by an interaction which gets stronger the farther apart the quarks are. Gravity and electromagnetism both get weaker at longer distances, but we are all familiar with a force that gets stronger at greater distances. If you stretch an ordinary elastic band it will resist your efforts more and more strongly as it stretches, up to the point where it breaks. Quarks behave as if they are held to their immediate neighbors by floppy elastic bands and are rattling around loose inside the nucleus, but they are tugged back sharply if they move away from one another. The analogy can even be carried over to the snapping of elastic bands. If enough energy is put into moving one of the quarks — for example, if it is struck by a fast-moving particle from outside in an accelerator experiment — the interaction with its neighbors will indeed be broken. But, in line with Einstein's famous equation $E = mc^2$, this happens only if there is enough spare energy (E) to make two new quarks (each with mass m). All the extra energy goes into making those new quarks, one on each side of the break, so you still do not detect an isolated quark.

This business of making particles out of pure energy (if you like, $m = E/c^2$ rather than $E = mc^2$) is itself crucially important to our understanding of the subatomic world. In particle colliders, beams of energetic particles are smashed head-on into one another or into stationary targets. When this happens, the fast-moving particles are brought to a halt, and the energy of motion that had been put into the particles is released in the form of a shower of new particles. These are *not* particles that were present in any sense "inside" the original particles and have been knocked out by the collision; they are new particles that have literally been made out of pure energy. Most of the particles made in this way are unstable and break down into less massive particles, eventually into the familiar protons, neutrons, and electrons. But the way they break down provides clues to their own internal structure, and this has led to the improvement of the standard model.

The first step was to find a model to describe the strong interaction. The interaction which confines quarks inside nucleons is now regarded as the true strong interaction; the force between nucleons, the original strong interaction, is seen as a weaker trace of this real strong interaction, which leaks out of nucleons to affect their neighbors. Once the evidence in support of the quark model became convincing, physicists were quickly able to come up with a model of the strong interaction operating between quarks because back in the 1940s they had developed an exquisitely precise model of the way electrically charged particles such as electrons and protons affect one another through the electromagnetic interaction.

This model was based on the idea of a field, which is familiar in the context of a magnetic field—an influence spreading out through space from some source. In the case of a magnetic field, you can even get a visual image of what is going on by placing a bar magnet underneath a sheet of paper, sprinkling iron filings on top of the paper, and gently tapping the paper to make the filings arrange themselves in curving patterns following the "lines of force" of the magnetic field. Because modern field theory incorporates ideas from quantum physics, it is called quantum field theory. The specific piece of quantum physics that goes into the theory of the electromagnetic interaction is that light—*light* here meaning not just visible light but any form of electromagnetic radiation, including radio waves, X-rays, and so on— comes in the form of quanta called photons. In the language of quantum physics, the photons are called field quanta and are regarded as bits of the field that have been "excited" by an input of energy.

In the 1930s physicists developed the idea that the electromagnetic interaction could be described in terms of an exchange of photons between charged particles. The early version of this model made predictions about the behavior of charged particles that were close to the properties observed in experiments but that didn't quite match up to measurements of what was going on in interactions between charged particles. In the 1940s, however, these discrepancies were resolved, and the modern theory of quantum electrodynamics (QED) emerged by taking on board one of the strangest aspects of the quantum world, uncertainty.

Quantum uncertainty is actually a very precise thing. The idea was developed by the German physicist Werner Heisenberg in the late 1920s, initially in terms of two familiar properties of particles: their position and their momentum, which is a measure of the direction and speed they are moving. In the everyday world, we are used to the idea that we can, in principle, measure both the position and the momentum of an object (such as the archetypal billiard ball) at the same time. We know *simultaneously* both where the object is and where it is going. Heisenberg realized that quantum entities such as electrons and photons do not behave like this — something that becomes obvious in hindsight if you think a little about wave-particle duality. Position is a typical property of a particle, but waves do not have a precise position in space. If a quantum entity has (or behaves as if it has) aspects of both particle and wave in its nature, it is no surprise to find that it can never be located precisely at a point.

Heisenberg found that the amount of uncertainty in the position of a quantum entity (the uncertainty about where it is) is related to the amount of uncertainty in its motion (the uncertainty about where it is going) in such a way that the more precisely the position is confined the less certainty there is about the momentum, and vice versa. The two uncertainties are linked mathematically by an equation now known as Heisenberg's uncertainty relation. And the crucial point to take on board is that this uncertainty is not a result of human clumsiness or the inadequacy of experiments we build to measure things like electrons. It is built into the nature of the quantum world. An electron literally does not have a precise position and a precise momentum. An electron confined within an atom, for example, is fairly precisely located in space, but its momentum is constantly changing as it moves around in the atom's electron cloud; an electron moving through space like a wave may have a very precise momentum, but it does not exist at a point anywhere along the wave.

Strange though all this is, it is not the end of the story. The same kind of quantum uncertainty carries over into other pairs of properties in the quantum world, and one of these pairs is energy and time. Heisenberg's uncertainty relation combined with Einstein's special theory of relativity (which is all about space and time) tells us that if you take a certain volume of seemingly empty space and monitor it for a certain time, you cannot be sure how much energy it contains. It's not just "you" who cannot be sure; as with position and momentum, nature itself does not know. If you monitor it over a long period of time, you can be certain that the space is empty (or nearly empty). But the shorter the period of time involved, the less certain you can be about the amount of energy inside the volume. For a short-enough interval of time, energy can fill the volume, provided it disappears again within the time limit set by the uncertainty relation.

This energy could take the form of photons, which appear out of nothing at all and promptly disappear again. Or it could even take the form of particles like electrons, provided that they exist only for the tiny flicker of time allowed by the uncertainty relation. Such short-lived entities are known as "virtual" particles, and the whole process is called a "fluctuation of the vacuum." In this model, empty space, or the vacuum, is seen as seething with activity on the quantum scale. In particular, a charged particle such as an electron is embedded in a sea of virtual particles and photons, and even in their short lifetimes the particles interact with the electron. When QED was adapted to take account of the presence of this sea of virtual particles, it gave predictions that precisely matched the properties of charged particles measured in experiments. In fact, the experiments and the model match to a precision of one part in ten billion, or 0.00000001 percent. The only reason the accuracy isn't even better is that experiments good enough to test the model to even greater precision have not yet been devised. This is the most accurate agreement between theory and experiment for any scientific model tested on Earth; even Newton's law of gravity hasn't been tested to such precision. By that measure, QED is the most successful model in the whole of science. And the agreement is only that good if the effects of quantum uncertainty — a seething vacuum and virtual particles — are included. The whole model passes the test.

It is hardly surprising that when physicists wanted to develop a model of the interaction between quarks — the strong interaction — they took QED as the template and tried to come up with a similar quantum field theory. In this model, the field quanta that are responsible for conveying the strong interaction are called gluons, because they glue quarks together. Just as photons are associated with electric charge, gluons are associated with another kind of charge, which is called color but has nothing to do with color in the everyday sense of the word. Whereas electric charge comes in only two varieties, positive and negative, color charge comes in three varieties, "red," "blue," and "green." To make the model of the strong interaction work, you need eight different kinds of field quanta, whereas with electromagnetism you need only one kind, the photon. In addition, unlike the photon these gluons have mass.

The model of the strong interaction based on QED is called quantum chromodynamics, or QCD, because it uses the names of colors in this way. Because of complications caused by the larger number of kinds of field quanta and the fact that they have mass, QCD does not make predictions that match the results of experiments as precisely as those of QED, a hint that the standard model is not the last word in physics. But it is the best model we have for what is going on inside things like protons and neutrons.

Field quanta such as photons and gluons are collectively known as bosons (in honor of the Indian physicist Satyendra Bose), while the kind of entities we are used to thinking of as particles, such as electrons and quarks, are known as fermions (after the Italian physicist Enrico Fermi). Just as bosons can be thought of as field quanta, so fermions are regarded as the quanta associated with "matter fields" that fill all of space, blurring the distinction between "particles" and "forces" further. But there *are* differences. The main difference between the two families is that bosons can be created out of pure energy without limit — every time you turn a light on, billions and billions of newly created photons flood into the room. But the total number of fermions in the Universe has stayed the same, as far as we can tell, since the Big Bang. The only way you can make a "new" fermion, such as an electron, out of energy is by making a mirror-image antiparticle (in this particular case, a positron) at the same time. The mirror-image particle has opposite quantum properties (including, in this case, positive electric charge instead of negative electric charge), so the two cancel each other out for the purpose of counting fermions, with one negative and one positive adding up to nothing. We'll have more to say about antimatter later.

So far we've identified three different fermions — the electron, the up quark, and the down quark. We've also identified three different kinds of interaction — gravity, electromagnetism, and the strong interaction. But there is still one more fermion and one more interaction to add to the list. These additions to the standard model are required to explain a phenomenon first observed in the nineteenth century but only satisfactorily described mathematically in the 1960s.

It is called beta decay and involves processes in which electrons (which used to be known as beta rays) are ejected from electrons. It took physicists so long to understand what was going on because the nature of the phenomenon seemed to change as they probed deeper into the structure of the atom.

It is no surprise that *atoms* can eject electrons, since all atoms contain electrons. But experiments showed that the electrons involved in beta decay actually come from the *nuclei* of atoms, and nuclei do not contain electrons, only neutrons and protons. The experimenters next discovered that in beta decay a neutron spits out an electron and converts itself into a proton. There is no change in the total electric charge in the Universe since the positive and negative cancel each other out, but an extra fermion seemed to have been created. In addition, in order to balance the energy and momentum of the ejected electron, it seemed that there ought to be an invisible particle flying away from the decaying neutron in the opposite direction. Both puzzles were resolved in the early 1930s by the suggestion that when an electron is made out of energy in beta decay, a partner fermion called a neutrino (strictly speaking, to balance the number of fermions, an *anti*neutrino) is also manufactured. Since neutrinos have no charge and only a very tiny mass, it took until the 1950s for this speculation (made by the Austrian physicist Wolfgang Pauli) to be confirmed by experiment, but confirmed it was. Yet even then it was clear that neither the electron nor the neutrino exist "inside" a neutron; in beta decay the internal structure of a neutron is rearranged to release energy, in the form of these two particles, and convert the neutron into a proton.

This is now understood in terms of quarks. A neutron contains two down quarks and one up quark, while a proton contains two up quarks and one down quark. The down quark has a negative electric charge ⅓ the size of the charge on an electron, and the up quark has a positive electric charge ⅔ the size of the charge on an electron. So if a down quark is converted into an up quark, one negative unit of charge has to be carried away, and the absence of this negative charge — in a splendid example of two negatives making a positive — leaves an overall balance of one unit of positive charge behind; the neutron becomes a proton. The negative charge is carried away by an electron, and some of the excess energy is carried away by an antineutrino. Both the total number of fermions and the overall amount of electric charge in the Universe remain the same. Since the mass of a down quark is more than the mass of an up quark, and mass is equivalent to energy, everything balances nicely. As long, that is, as there is an extra kind of interaction going on between the particles involved.

This "new" interaction became known as the weak interaction (because it is not as strong as the strong interaction), and it was developed to provide an insight into processes that involve radioactive decay (when nuclei split apart) and nuclear fusion (when nuclei join together to make more complex nuclei, as happens inside stars). In order to match the data from experiments, the weak interaction requires the existence of three kinds of bosons, the W^+ and W^-, which each carry the appropriate unit of electric charge, and the Z, which is electrically neutral. This makes it easier to deal with mathematically than QCD, but more complicated than QED. Although the theory of the weak interaction now describes much more than simple beta decay, to complete that story the modern picture of beta decay envisages a down quark emitting energy in the form of a W^- boson and converting itself into an up quark, and then, after a very short time, the energy of the W^- boson converting itself into mass in the form of an electron and an antineutrino.

Like gluons, the W and Z particles have mass, and their masses can be predicted from the model. It was one of the greatest triumphs of the standard model that when those particles were detected, in the early 1980s, at the CERN laboratory near Geneva, they were found to have exactly the masses predicted by the model. By then, though, the standard model had become more complicated in some ways, though simpler in others.

The essence of the standard model describes the familiar physical world in terms of just four particles* and four interactions. The four particles are the electron and the neutrino (collectively known as leptons) and the up and down quarks. The four interactions are gravity, electromagnetism, and the weak and strong nuclear interactions. These are all physicists need to explain every natural phenomenon on Earth, as well as the workings of the Sun and all the stars we can see in the sky. But to their surprise, they turned out not to be enough to explain everything observed in the unnatural conditions of high-energy processes going on in particle accelerators.

There really do seem to be just four interactions operating in the Universe. The surprise is that the particle world is not only duplicated but triplicated at high energies. If enough energy is available, it is possible to manufacture short-lived massive counterparts to all the four basic particles in two further generations. First, there is a heavy counterpart to the electron called the muon, with its associated muon neutrino, and two heavier quarks, called charm and strange; there is also an even heavier "electron" called the tau, with its own tau neutrino, and two very heavy quarks, called top and bottom. Subtle ex-

---

*We mean four *kinds of* particles, of course, but it would be pedantic to spell that out every time.

periments at CERN have proved beyond all reasonable doubt that this is the end of the story; no matter how much energy you put into a particle collision you will never make a fourth generation of particles.

When these heavier particles are manufactured in accelerators, they promptly decay, eventually degenerating into the familiar particles of the first generation. So in the world today they are of only academic interest. But in the energetic conditions of the early Universe, they would have been manufactured in profusion, and they had an effect on how the Universe evolved. Nobody knows why the Universe should be so profligate as to allow for the existence of these heavier versions of the basic set of four fundamental particles. It is just another sign that the standard model is not the last word in physics.

But don't despair. Even while the standard model was incorporating this unwelcome addition into the particle zoo, it was eliminating one of the interactions from the scheme and pointing the way toward eliminating others.

Apart from the mass and charge of the particles, the equations that describe the W and Z bosons as the particles associated with the weak field are very similar to the equations that describe the photon as the particle associated with the electromagnetic field. And electric charge is already described by Maxwell's equations of electromagnetism. In the 1960s, physicists realized that if they could find a way to add mass to photons they would have a single set of equations to describe both the electromagnetic field and the weak field — they would have combined the two fields into one "electroweak" interaction. The search for such a unification of the two interactions took theorists up several blind alleys before they hit upon a satisfactory model — "satisfactory," as always, meaning that it makes predictions which match the results of experiments. The model they came up with, now a key component of the standard model, was actually developed from an attempt by the

British physicist Peter Higgs, working at CERN, to find a model for the strong interaction. As is usually the case, many people were involved in developing the model, but it was Higgs who gave his name to the phenomenon.

The idea that Higgs came up with is that all particles are intrinsically massless, but a previously unsuspected "new" field, which interacts with particles to give them mass, fills the whole Universe. This field is now known as the Higgs field. An easy way to envision what is going on is to imagine the way the behavior of a spacecraft would be altered if space were actually filled with an invisible gas, like air. In empty space, if the rocket motors of a space probe are used to provide a steady push on the probe, it will accelerate at a steady rate as long as the motor keeps firing. But if the probe were moving through a completely uniform sea of gas, then when the motors were fired at the same steady rate it would not accelerate as quickly because of the drag caused by the gas. The effect would be the same as it would be if the probe were heavier (more massive) than it really was. In an analogous way, massless particles moving through the Higgs field encounter a "drag" which seems to give them mass, with the exact mass depending on the nature of the individual particle and the strength of the influence it feels from the Higgs field.

This model makes predictions about the masses of the W and Z particles, and in 1984 experiments at CERN reached the energies where particles with the required mass could be manufactured in line with $E = mc^2$. The particles were found exactly as predicted and with exactly the right masses. This is one of the greatest successes of the standard model. But the model also makes one key prediction which has not yet been tested.

According to the model the Higgs field, like all fields, must have a particle associated with it — the Higgs boson. This particle is far too massive to have been manufactured in any experiments yet carried out on Earth. But in 2007 a new accelerator, called the Large Hadron Collider (LHC), is due to begin operating at CERN. Some idea of the effort that is required to probe the nature of the Universe at this level is given by the size and cost of the LHC. Buried a hundred meters underground in a circular tunnel 27 kilometers in circumference carved through solid rock, the LHC will take beams of protons that have been preaccelerated to high energy in the existing accelerators at CERN and send them opposite ways around the ring to smash head-on at a collision energy of 14 tera electron volts (TeV; 1 TeV is a thousand billion electron volts). This is about the same as the energy of motion (kinetic energy) of a flying mosquito — but packed into a volume a thousand billion times smaller than a mosquito. It is enough energy to make a thousand protons out of pure energy. The LHC will also be able to collide beams of lead nuclei into one another at energies of a little over a thousand TeV. It will use 1,296 superconducting magnets and 2,500 other magnets to guide and accelerate the particle beams, and comes in at a cost, in round numbers, of about 5 billion Euros (just under £3.5 billion or $6 billion). That's the price we have to pay if we want to test the standard model. If the standard model is indeed correct, the LHC should start to manufacture Higgs particles soon after it becomes fully operational. If the Higgs boson is found as predicted, it will make the standard model even more secure, and Peter Higgs will be sure to get a Nobel Prize; if it does not turn up at the predicted mass, this will point the way toward a better model of the subatomic world.

So the standard model of particle physics, what we think we *know*, now incorporates four basic particles in two pairs (electron and proton, up and down quarks), which are repeated (for reasons unknown) in two further generations. It also incorporates three interactions (gravity, the electroweak interaction, and the strong interaction), plus the Higgs field. This package explains everything on Earth, as well as how the stars work. But physicists want to do more. They want to explain where the Universe came from, and how stars and planets came into existence. As we shall see in Chapter 3, there is good evidence that the Universe began in a hot fireball involving energies far greater than anything that can be achieved in our experiments. So in order to try to understand where the world came from and, ultimately, where *we* came from, theorists have to go beyond the standard model to things we *think* we know.

2

Is There a Theory of Everything?

The success of electroweak unification encouraged physicists to attempt to unify the electroweak interaction with the strong interaction, and to dream of incorporating gravity into the package at a later date. Models which attempt to describe the electroweak and strong interactions in one mathematical package are usually called grand unified theories, or GUTs; a theory incorporating gravity as well would be a theory of everything, or TOE. In both cases, with our present state of knowledge the word *theory* might be better replaced by *hypothesis* or *model*; but the acronyms would not be as convenient.

Although we introduced the idea of the weak field in terms of neutron decay, there is another side to this kind of interaction which helps to show how the grand unified theories are constructed. Instead of a neutron spitting out an electron and an antineutrino and converting itself into a proton, a neutrino from outside can interact with the neutron through the weak interaction (strictly speaking, it interacts with the down quark inside the neutron). In the process, the neutrino is converted into an electron (and the neutron into a proton). In other words the weak interaction can change one kind of lepton into another kind of lepton. In QCD, when gluons move between quarks, they carry color charge with them, and so they can change the colors (we should say, the color charge) of the quarks. In other words, the strong interaction can change one kind of quark into another kind of quark—a trick which the weak interaction also does, in a different way, when it changes a down quark into an up quark inside a neutron. So it is possible to change one kind of particle (one kind of fermion) into another.

In the wake of the successes of the electroweak unification and QCD, the question physicists then asked themselves was whether it was possible to find a model in which one kind of charge carrier (one kind of boson) could be changed into another. Their attempts to answer this question led to the development of GUTs. Yet the fact that we still have to talk about a variety of possible grand unified theories, rather than a single definitive theory, shows that this work is still incomplete. There was, however, one key implication to all such models (as they should be called): bosons which can change one *kind* of charge into another *kind* of charge would be able to convert leptons into quarks, and vice versa. This came as no great surprise, since the discovery that quarks and leptons come in three generations, with a pair of leptons and a pair of quarks in each generation, already hinted at a fundamental link between the two kinds of fermions. It meant, though, that just as neutrons can interact with bosons of the weak field to decay into protons, protons should be able to interact with bosons of the unification field to convert themselves into leptons. In other words, even protons should decay, as quarks inside them are converted into leptons. This is a firm, testable prediction of this entire family of GUTs.

The hypothetical bosons of the unification field are dubbed X and Y particles. They have never been detected, but this is no surprise because the models tell us that they must have enormous masses, by particle standards — typically a million billion times (10^{15} times) the mass of a proton, so that to make them would require an accelerator capable of reaching energies 1,000 billion (10^{12}) times greater than those at which the W and Z particles were found.* The various X and Y particles would also have electric charges equivalent to plus or minus ⅔ of the charge on an electron, or plus or minus ⅓ of the charge on an electron; they also come with different color charges. A proton could decay today only if it could "borrow" an appropriate amount of energy from quantum uncertainty, in the form of a fluctuation of the vacuum, to make such a boson for a brief instant.

If it could do so, one form of proton decay would proceed like this. An up quark inside the proton (charge ⅔) would convert itself into an anti–up quark (charge $-⅔$) and release an X boson (even though the mass of the boson is far greater than the mass of the quark!). The X boson (charge ⅔) would also interact with a neighboring down quark, converting it into a positron. That would leave the anti–up quark partnered by the one remaining quark from the proton, an up. The up quark and its antimatter counterpart, the anti–up quark, would form a temporary bound state (called a pion) but would quickly annihilate each other, leaving the positron and a burst of

---

*The units used by particle physicists to measure mass are called electron volts (eV). The mass of a proton is roughly one giga (billion) electron volts, or 1 GeV. So the mass of an X particle is about 10^{15} GeV. In fact, physicists measure *both* energy and mass in electron volts since, as Einstein discovered, mass and energy are equivalent to each other. The factor of c^2 which appears in Einstein's most famous equation is taken as read.

energetic electromagnetic radiation (photons) as the only trace of the original proton. But everything would depend on borrowing the energy from the vacuum to make a virtual X boson for long enough for the interaction to occur. The virtual X boson exists for such a short time that it can travel no farther than 10^{-29} centimeters before it disappears, so in order for it to make its influence felt, two quarks have to come at least as close as 10^{-29} centimeters to each other.

The models tell us how likely this is. According to calculations based on the simplest versions of those models, the chance that enough energy would appear out of nothing at all, as a fluctuation of the vacuum within an individual proton, is so small that if you had a box full of protons and waited for rather more than 10^{30} years (that's roughly 100 billion billion times the age of the Universe) just half of them would decay. It doesn't matter how many you started with. If you started with a thousand, after all that time five hundred would have decayed; if you started with a billion, after all that time half a billion would have decayed. For obvious reasons, this characteristic timescale is known as the half-life of the proton, and there are different characteristic half-lives for all such decay processes. (For example, the half-life of a neutron outside the nucleus — things are different inside a nucleus — is 10.5 minutes.) Clearly, for an individual proton, the chances of such a decay happening in a human lifetime are vanishingly small. But there are a lot of protons in the Universe (your body, for example, contains about 10^{29} protons). If you watch a lot of protons for a fairly short time, you ought to see at least a few of them decay. If you watch a lump of matter containing about ten tonnes of stuff with a half-life of 10^{30} years you ought to see about five decays per year. And this is true of anything — water, iron, sausages, anything containing protons. Scientists have carried out such experiments (often using large tanks of water, since it is easy to handle), looking for the tell-tale positrons produced by such decay. So far, no trace of proton decay has been seen, and this tells us that the half-life of the proton is at least 5×10^{32} years (a five followed by thirty-two zeroes years). This is longer than the half-life predicted by the original GUTs. But that is actually good news, since attempts to improve those models for other reasons led to changes in the calculations which imply a longer half-life for the proton.

One of the improvements has to do with the way the strength of a force seems to change as you approach a particle. When we talked about the strengths of the interactions before, we were referring to the strength that would be measured at a considerable distance from the particle involved, compared with its own size. Since a "considerable distance" for an electron is still tiny by human standards, this is a good first approximation. But the cloud of virtual charged particles that surrounds an electron partially shields its bare intrinsic charge from the outside world. This shielding makes the electromagnetic interaction seem weaker from a distance than it would if you could get really close to an electron. So for an electron or any other electrically charged particle, on very small scales the strength of the electromagnetic interaction seems to get stronger as you approach the particle. On the other hand, the way quarks and gluons interact with virtual particles makes the strong interaction get weaker as you approach a particle that feels the strong interaction. Dealing with the electroweak interaction is a little more complicated than the case of electromagnetism, because of the effect of the masses of the W and Z particles. When those are taken into account, the "bare" strength of the interaction is in between those of the strong interaction and the electromagnetic interaction, and it also gets weaker at closer distances.

If the standard model were perfect, all these interactions would seem to have the same strength on a certain distance scale. As we have mentioned, probing shorter distances involves accelerating beams of particles to higher energies, so this is equivalent to saying that at a certain energy the strong, weak, and electromagnetic interactions should show the same strength. The distance involved is indeed extremely tiny—about 10^{-29} centimeters. It's almost impossible to get your head around what such a tiny number means, but if a typical atomic nucleus were expanded to a become a ball a kilometer across, an entity that started out 10^{-29} centimeters across would become the size of a nucleus. The equivalent energy is about 10^{15} times the mass energy of a proton, way beyond the levels we can ever hope to achieve in our experiments. It is no coincidence, though, that this energy corresponds to the masses of the X and Y bosons. Another way of thinking about all of this is in terms of the energy available to make new particles. If there is enough energy to make X and Y bosons, then there is certainly enough to make W and Z particles as well, and all the gluons you could ever desire. So all the virtual particles are promoted into reality—they are not virtual any more but use the available energy to become real particles, which no longer have to form a tight-knit cloud around particles like electrons, and can wander off on their own. So the shielding effect disappears, and all we see are bare charges (electric charges, color charges, and so on).

The weakness of the standard model and its extensions described so far is that although it predicts the energies at which all these processes occur, it doesn't predict exactly the same energy for the unification of all three interactions. It says that the electromagnetic and weak interactions merge into the electroweak interaction at one energy, the electromagnetic and the strong interactions merge at a slightly different energy, and the weak and strong interactions merge at a third,

slightly different energy. But the three unifications can be pulled together to meet at a single point by a family of GUTs (models), which
go by the name of supersymmetry, or SUSY.

We have already encountered symmetries in which one kind of
fermion is changed into another kind of fermion, and one kind of
boson is changed into another kind of boson. The suggestion made in
the mid-1970s by Julian Weiss, in Germany, and Bruno Zumino, in
California, was that fermions and bosons might be related to one
another by another kind of symmetric process — supersymmetry —
which could convert fermions into bosons and vice versa.

To everyday common sense, this seems ridiculous. How could
matter be turned into force, or force into matter? But the quantum
world often conflicts with everyday common sense, and we've already
come across something just as bizarre. In the quantum world, waves
and particles are interchangeable, two facets of the same thing, even
though in our everyday world we think of things like electrons as
particles and of forces like electromagnetism in terms of waves. So it
isn't that stunning a leap to think of force carriers being interchangeable with matter particles. In quantum language, supersymmetry
states that fermions can be converted into bosons and bosons into
fermions. But you can't turn any old fermion into any old boson; each
kind of particle has to be paired up with its own kind of supersymmetric partner.

But where are their partners? We already know that leptons and
quarks come in related versions, so that the electron and its neutrino
"belong" with the up and down quarks. But none of the known matter
particles belongs in the appropriate supersymmetric sense to any of the
known force carriers, and none of the known bosons belongs to any of
the fermions. The theorists are unfazed by this, and suggest (with the
mathematics to back them up) that every kind of known fermion

(such as an electron) should have a supersymmetric partner (in the case of the electron, it's called a selectron), which has never been seen, and each type of boson (for example, the photon) should have a fermionic counterpart (in the case of the photon, the photino), which has also never been seen. These hypothetical entities are collectively called SUSY particles. The reason they have never been seen, the argument runs, is that first of all they have large masses (so they have not been made in any accelerator experiments on Earth), and second, they are unstable, so they promptly decay into a mixture of the familiar kinds of fermions and bosons and lighter SUSY particles. There is one exception, if the idea of supersymmetry is correct. The lightest of these "supersymmetric partners" (probably the photino) would have to be stable, because there could be nothing lighter for it to decay into.

The reason this package of ideas is taken seriously is that the changes to the standard model required to make room for the simplest version of supersymmetry (known as minimal supersymmetry) shift the predicted convergence of the electromagnetic, weak, and strong force couplings so that they all meet precisely at a point, at an energy of about 10^{16} GeV rather than 10^{15} GeV. The addition of supersymmetry to the equations also changes the prediction for the half-life of the proton, raising it above the level reached so far by experiments.

So there is no evidence that supersymmetry is wrong; and there may soon be some evidence that it is right. One of the reasons why physicists are excited about the prospect of the Large Hadron Collider is that it should be able to manufacture supersymmetric partners with masses of a few thousand times the mass of a proton (a few thousand GeV). It should also manufacture Higgs bosons. If the LHC *doesn't* manufacture Higgs particles, it will be a major surprise, and require a rethinking of the whole idea of GUTs (which would nevertheless be a delight for physicists, as it would give them something to think about). But one reason why that seems unlikely is that there has already been one triumph for the GUTs. And the origin of that evidence in their favor shows how particle physicists are increasingly turning to astronomy to find ways to test their models.

Another prediction from minimal supersymmetry is that neutrinos should have a tiny mass. In the standard model without SUSY, neutrinos should be completely massless, like the photon. This had been a puzzle since the late 1960s, when an experiment designed to monitor the flow of neutrinos from the Sun had begun to show too few neutrinos arriving at the Earth. Electron neutrinos are produced in profusion by the nuclear reactions that go on at the heart of the Sun and keep it shining, and the number of these neutrinos flooding past (and through) the Earth can be predicted from standard models of nuclear physics and astrophysics. But when Ray Davis and his colleagues, working at the Homestake Gold Mine in Lead, South Dakota, set up an experiment to monitor this flux of neutrons, they found about a third of the predicted number. Assuming the nuclear physics and astrophysics models are correct — and there is a wealth of independent evidence that they are — one possible explanation is that the electron neutrinos were changing into other kinds of neutrinos on their way to us. This process is called an oscillation, because the electron neutrinos would change into muon and tau neutrinos and back again as they travel through space, or mixing, because the original electron neutrinos are mixed into all three varieties.

Because there are three kinds of neutrino all together, and the mixing is even, this process would naturally ensure that one-third of the original flux of electron neutrinos would be noticed by the Davis detector, which could not "see" the other kinds. But such an oscillation would work only if the neutrinos had mass. That this was indeed the case was a dramatic discovery in the 1970s, and a new kind of development in physics. *Astronomical* discoveries were telling physicists about the properties of the tiniest particles known. Since these pioneering experiments, other studies of solar neutrinos and direct observations of neutrino oscillations in experiments on Earth have confirmed that the astronomers were right. Neutrinos do have mass (less than one-tenth of an electron volt; for comparison, an electron has a mass of *511,000* eV),˙ and they do oscillate. In the wake of these discoveries, as we shall see shortly, the links between astronomy and particle physics have become ever stronger.

Overall, the combination of grand unified theories with supersymmetry (SUSY GUTs) looks extremely promising and makes predictions that will be tested before the end of the first decade of the twenty-first century. If all goes well with those tests, the next step will be to find ways to incorporate gravity into the package, making a real theory (model) of everything, or TOE. In general terms, the way to achieve this would be to describe gravitational interactions in terms of the exchange of particles called gravitons, and to bring gravity into the SUSY fold by postulating the existence of a supersymmetric partner, the gravitino. Gravitons arise naturally in any quantum theory of gravity, just as photons arise naturally in a quantum theory of electromagnetism, but it is the addition of the gravitino that brings gravity into the supersymmetric description of the world. Such variations on the unification theme go by the generic name of "supergravity"; but they are more speculative than anything we have yet discussed and have yet to be tested by experiments. In any quantum theory of gravity, including supergravity, gravitons would have to be massless, like photons, in order to give gravity the same long-range influence as electromagnetism. But, unlike photons, gravitons can (according to the models) interact with one another, which makes the calculations vastly more difficult.

Just as electromagnetism can be described either in quantum terms (photons) or in classical terms (waves), so gravity can be described in terms of gravitons or in terms of gravitational waves—part of the wave-particle duality that lies at the heart of quantum theory. There is much more prospect of studying gravitational waves in the near future than there is of testing supergravity models in experiments. Because gravity is such a weak force, extremely sensitive detectors will be required to identify the waves associated with gravitons. But such detectors are now under construction and may detect gravitational radiation within the next few years.

Gravitational waves are a prediction of Einstein's general theory of relativity, which treats space (strictly speaking, space-time) as an elastic entity that is distorted by the presence of matter (and is a "classical" theory, meaning that it does not involve quantum processes). Think of empty space-time as being like a stretched, flat rubber sheet; marbles rolled across the sheet will travel in straight lines. But if you place a heavy object, like a bowling ball, on the sheet, it makes a dent; now, marbles rolled near the heavy object follow curved paths around the dent. With the appropriate mathematics put into the picture, this explains precisely why, and by how much, light is deflected when it passes near the Sun—a prediction of Einstein's theory that was famously tested and confirmed by observations made during an eclipse of the Sun in 1919.

But if you go one stage further, you can imagine the bowling ball being bounced up and down on the rubber sheet and making ripples in the fabric. Matter that jiggles in the right way in the Universe at large will produce ripples in space-time, according to Einstein's equations, and those ripples should be detectable as waves in three-dimensional space. The effect is actually rather small, because gravity is such a very weak force compared with the other three forces of nature — which is just as well, or any ordered structures in the Universe (including ourselves) would be shaken to bits by the gravitational radiation criss-crossing the Universe. But the most dramatic "jiggles" produced by matter, such as the collapse of a star into a black hole, should produce ripples in space big enough for the new generation of instruments to detect. Clear evidence of the existence of gravitational radiation has already come from astronomical studies of the behavior of a system known as "the binary pulsar," in which two neutron stars orbit around each other. So we know for sure that Einstein was right and such radiation exists. But physicists are still eager to detect gravitational waves here on Earth, and they should do so very soon.

We don't intend to go into great detail about all the experiments we shall mention in this book; the results are more important than the details of how they were obtained. But it is, perhaps, worth picking out one example to represent the kind of international collaboration involving large teams of researchers that typifies the way science is being done in the first decades of the twenty-first century.

There are four of these new gravitational wave experiments now running. The biggest (LIGO) is in the United States; there is also one in Japan (TAMA), a joint French-Italian experiment (VIRGO), and the one we shall describe in some detail, a joint British-German detector known as GEO600. These experiments are not merely a competitive quadruplication of an effort you might think ought to be devoted to one single global project. You need at least two detectors just to be sure that you have indeed detected gravity waves, since they will both record those vibration at the same time, confirming that the waves are not a local disturbance caused by a passing truck or a landslide. And in order to identify the position in the sky of the source of the waves, and other detailed properties of the radiation, you need a minimum of four detectors. They all operate on similar principles, but in some ways GEO600 is the most sophisticated, because severe financial constraints have forced the experimenters to use extreme ingenuity and develop new technology to achieve their objectives. The same financial constraints forced the British and German teams into what has proved a happy collaboration at the end of the 1980s, since neither Britain nor Germany could afford its own gravitational wave detector. This is very much the way big science is going in the twenty-first century, and as we shall see, it is extremely rare for cutting-edge research nowadays to be carried out by a single country (let alone a single research group at a single university) working in isolation. The days of the lone genius — a Newton or an Einstein — are long gone.

The "GEO" bit of the project's name comes from Gravitational European Observatory — the more natural order European Gravitational Observatory was felt to give an acronym with the wrong public image, even if the experimenters do have a high opinion of themselves. The "600" refers to the size of the experiment, which consists of two arms, each 600 meters long, at right angles to each other in an L shape. The size of the arms was dictated by the space available, located on agricultural land just south of Hanover that was owned by the Bavarian state and operated by the University of Hanover as an agricultural research center. The arms are built alongside farm roads dividing the fields, among crops of cereals and fruit. In fact, one of the arms sticks out past the boundary of the agricultural research area, onto the land of an adjoining farm, for a distance of 27 meters. The GEO600 budget pays the farmer 270 marks a year in rent for the privilege.

Each arm of the experiment houses a tube just 60 centimeters in diameter made of corrugated metal just 0.8 millimeters thick. Inside the tubes there is a vacuum as empty as space, with mirrors suspended in the vacuum to reflect light from a laser beam shone along the tube.

Each mirror weighs six kilograms and is suspended from four "wires" made of glass and only 200 millionths of a meter thick. The whole system is so delicately poised that by analyzing the light signals bounced off the mirrors by the lasers the researchers will eventually be able to measure changes in the length of each arm of the detector of less than 10^{-18} meters (that is, less than a billionth of a billionth of a meter, or one millionth of the diameter of a proton). At the end of 2004, Jim Hough, the head of the British team at Glasgow University, told me that test runs had reached a sensitivity about a factor of ten away from this target ("only" an accuracy of 10^{-17} meters), and that GEO600 should be working at its designed sensitivity by the end of 2006.

Gravitational waves should, according to the general theory of relativity, produce a distinctive "signature" as they pass through the experiment, first stretching one arm by roughly this amount and simultaneously squeezing the other arm, then reversing the process. It's like a quake in space-time that simultaneously makes you both taller and slimmer, then reverses to make you shorter and fatter. It is this distinctive pattern of events that makes it possible to measure such tiny changes. Even the GEO600 system cannot measure the length of one single evacuated tube to such precision, but by comparing the way laser beams in the two arms of the apparatus interfere with one another, the scientists can measure the relative changes in the two arms. If such a signal is found at GEO600 at the same time that similar changes are measured at LIGO or one of the other detectors, the researchers will know that they have observed the passage of a ripple in space through the Earth. Going beyond the excitement of the initial discovery — which could happen at any time now — future observations of such events will provide insight into the biggest explosions in the Universe, and perhaps into the nature of the Big Bang itself.

Hough says that there is a fifty-fifty chance that GEO600 will pick up such a signal by the year 2009. If it fails to do so, the next step will be to upgrade LIGO by installing detectors based on GEO600's pioneering designs in the bigger experiment. (LIGO has arms 4 kilometers long but less sophisticated detectors, giving it about the same sensitivity as GEO600 in its present incarnation.) Then, says Hough, he is "100 percent" sure that gravitational radiation will be detected in the second decade of the present century. One reason for his confidence is that whatever happens with the ground-based experiments, a space experiment known as LISA (Laser Inteferometric Space Antenna) is due to fly in 2012. This will consist of three spacecraft flying in formation in an orbit around the Sun, five million kilometers apart at the corners of a triangle. Laser beams linking the three space probes will be able to measure changes in their separation, caused by the action of gravitational waves squeezing and stretching space itself, to a precision of about one-hundred-thousandth of a millionth of a meter (10 picometers).

Meanwhile, regardless of the progress of the search for gravitational radiation, the search for a theory of everything received a boost in the mid-1980s when one class of models, which had not been set up with gravity in mind, turned out to include automatically bosons with all the right properties to be the carriers of the gravitational interaction. These "string" models are currently the hottest game in town when it comes to trying to find a theory of everything.

The idea of strings came partly from the intrinsic interest mathematical physicists have in playing with equations and partly from a very practical problem associated with all models that treat particles as points with no radius or volume. The problem is that in situations like the one which describes the electric force, say, as proportional to 1 divided by the square of the distance from an electron, if the electron has no size the distance can go all the way down to zero. Dividing by zero gives you infinity, and equations which give you infinite answers don't make sense. The way around this is a trick called renormalization, in which you essentially divide one infinity by another to get a sensible answer. This can be made to work satisfactorily in the standard model and QCD, but it is really a counsel of despair. Many eminent physicists, including Richard Feynman, regarded renormalization as a sign that the models are seriously flawed.

String theory regards the fundamental entities of which the physical world is made as extended objects — strings — rather than points. The strings can be open, with their ends free, or closed, making tiny loops. And they exist, according to the models, on even smaller scales than anything we have considered yet — insofar as the idea of length has any meaning on these scales, a string would be about 10^{-33} centimeters long. This is about one-hundredth of a billionth of a billionth (10^{-20}) of the radius of a proton; putting it another way, if a proton were 100 kilometers across, a string would be the size of a proton. There is absolutely no prospect of testing these ideas by detecting strings directly, so the string idea stands or falls by the predictions it makes about the nature of the world on the scale of things like protons. (But this may be an appropriate place to remind you that it doesn't matter whether the world "really is" made of tiny strings, only whether it behaves *as if* it were made of tiny strings.) And there are two things about string models which make them a hot topic today. The first is that one class of string models has no need of renormalization or, rather, the models seem to renormalize themselves automatically, with no help from the mathematicians; the infinities cancel out all by themselves.

The second point — far more important in the eyes of most physicists — is that without being asked, the string models include the graviton. This came as a complete surprise. The theorists who were playing with string theory in the 1980s had not gotten as far as thinking seriously about gravity (although the idea of a theory of everything was always in the back of their minds), and they were both baffled and annoyed when, in order to make their equations balance, their models required the existence of a particle that didn't fit into the requirements of the standard model plus GUTs. When the penny dropped, and they realized that this particle was the graviton, the whole subject took off. But the flight it took them on looks fanciful indeed to outsiders.

The price you have to pay for the success of string theory is to take on board the idea of extra dimensions of space, above and beyond the familiar three spatial dimensions (forward-back, up-down, left-right) plus the fourth dimension, time, of everyday life. Curiously, this idea goes back to the 1920s, when physicists knew about only two kinds of interaction, gravity and electromagnetism. For a brief time, until the nuclear interactions were identified, it looked as if adding a fifth dimension would provide a 1920s equivalent of a theory of everything by unifying these two interactions; but the idea was discarded when more interactions were discovered and only revived half a century later.

The idea jumps off from Einstein's general theory of relativity, which describes gravity in terms of distortions in the fabric of four-dimensional space-time. In 1919 Theodor Kaluza, a young German mathematician, wondered what Einstein's equations would look like if written out to describe distortions in a five-dimensional space-time. He had no reason to think that such equations would mean anything in terms of the physical world; he was exploring the possibility simply out of mathematical curiosity. To his own surprise, he found that the five-dimensional version of the general theory of relativity was made up of two sets of equations—the familiar equations of the general theory itself and an even more familiar set of equations exactly equivalent to Maxwell's equations of electromagnetism. In a nutshell, if gravity can be thought of as a ripple in four-dimensional space-time, electromagnetism can be thought of as a ripple in five-dimensional space-time. The idea was developed further, to include ideas from quantum theory, by a Swedish physicist, Oskar Klein, and became known as the Kaluza-Klein model. The math works perfectly; the only snag is that there is no trace of a fifth dimension (that is, a fourth *spatial* dimension) in the everyday world. But the physicists got around this by invoking a trick called compactification.

Compactification can best be understood by an example. A thin sheet of flexible material such as rubber is actually a three-dimensional object, but from a distance it looks two-dimensional because its thickness doesn't show. For the purpose of the example, we pretend that it really is a two-dimensional sheet. We can take this a stage further by rolling the sheet up to make a tube, with the edges stuck together. The two-dimensional sheet is wrapped around the third dimension, and if we look at it from even farther away it now looks like a one-dimensional line. But each "point" on the line is really a little circle, or loop, around the tube, and ripples in the second (what we might call the up-down) dimension can run up and down the tube even though we cannot see them—the ripples carry energy, so they affect the behavior of the whole line. One of the two dimensions of the sheet is hidden from us simply because it is too small to be seen, but it still makes its influence felt. In a similar way, in the original Kaluza-Klein model the fourth dimension of space could be envisaged by imagining that every point of four-dimensional space-time is really a little loop, only 10^{-32} centimeters across, bent around in a fifth dimension.

To some physicists at least, this seemed an acceptable price to pay to have one set of equations to describe every known interaction. In quantum terms, the Kaluza-Klein model is relatively simple because it has to deal with only two bosons — the graviton and the photon. But soon more interactions were known, with more complicated behavior. In order to include the strong and weak interactions with all their bosons in the package would require even more dimensions wrapped up in even more complicated ways, and this was simply too much to accept at the time, so the Kaluza-Klein model became no more than a curiosity while the standard model was being developed. But a later generation of mathematical physicists grew up more at ease with the multidimensional approach,* and by the 1980s it was clear that a drastically different approach was needed to go from the standard model to a theory of everything. That new approach combines the idea of strings with the idea of extra dimensions.

In its modern, twenty-first-century form, the tiny loops of string that we have already described are thought of as being wrapped around in a total of twenty-six dimensions. The various things we are used to thinking of as particles (electrons, gluons, and so on) correspond to different vibrations of the strings, which carry different amounts of energy, rather like the way different vibrations of a guitar string correspond to different musical notes. The fermions can be explained relatively simply, in terms of vibrations in ten dimensions running one way around the loops of string. Six of these dimensions are compactified, to leave the four familiar dimensions of space-time. The richness of the bosonic world, however, requires vibrations in twenty-six dimensions, which run the other way around the loops of

---

*This kind of thing is usual in science. At one time, it was said that the general theory of relativity could be understood by only three men; now it is taught to undergraduates. The revolutionary ideas of one generation become commonplace to the next.

string. Sixteen of these dimensions are required to account for the rich variety of bosons, and these dimensions compactify as a set so that they are in a sense "inside" the ten-dimensional strings. Nobody is quite sure exactly what this means, and theorists argue about whether these dimensions are "real." But from our point of view what matters is that bosons behave *as if* they had these extra dimensions associated with them. The other ten dimensions are the same as the dimensions in which fermionic vibrations occur. Six of these dimensions compactify so that the vibrating strings produce the appearance of particles moving in four-dimensional space-time. Because the model requires the existence of two different sets of vibrations occurring on a single kind of string, it is sometimes referred to as heterotic string theory.

There's an additional curiosity about this, which highlights our imperfect understanding of the "extra" sixteen dimensions. All the particles can actually be described in terms of eight of the sixteen compactified dimensions, leaving room for a duplicate set of particles. Nobody quite knows what this means, either, but some theorists have speculated that there might be a complete "shadow universe" made up of these duplicate particles, which shares our four-dimensional space-time but has no interaction with us, except possibly through gravity. A shadow person could walk right through you without your noticing. But we will leave further speculation along those lines to the science-fiction writers. The real progress with string theory in recent years has come from reinterpreting the other part of the model, the ten-dimensional component.

So far, we've talked about string theory, in the singular, as if there were just one model which fitted the bill. This is the optimistic language used by the string theorists themselves, but in the ten years from the mid-1980s to the mid-1990s it disguised an awkward fact.

There were (and are) actually five different models, variations on the string theme, each of which offered a subtly different interpretation of what is going on, but all of which involved six compactified dimensions of vibrating string moving in four-dimensional space-time (plus the extra sixteen bosonic dimensions that nobody properly understands). This wasn't as distressing to the physicists as you might think, though, because they were able to prove mathematically that these were the *only* possible models — they could dream up mathematical versions of other kinds of string models, but they could show that all such models were plagued by nonrenormalizable infinities and made no practical sense.

There was also a joker in the pack — a single version of another kind of model, called supergravity, which seemed to explain things as well as any of the five string models, but which required the existence of eleven dimensions rather than ten. It turned out that instead of being a freak, the fact that supergravity only worked in eleven dimensions was an important clue to what was going on.

Following a huge effort by many theorists in the early 1990s, in 1995 the American physicist Ed Witten pulled everything together by adding an extra dimension to string theory. He showed that all six of the candidates for a theory of everything were different aspects of a single master model, which he called M-theory. In the same way that electromagnetism and the weak interaction look like different things at low energies but are really separate manifestations of single electroweak interaction, the six candidates for a theory of everything are low-energy manifestations of a single M-theory that would only be apparent experimentally if we could generate energies equivalent to those of the strong interaction. The price Witten had to pay for this success was to introduce an extra dimension of space to the string models, so that, like supergravity, they operated in eleven-dimensional space-time. Another tiny, compactified dimension might not seem like much of a step forward when you've already got six of them; but this "new" dimension in M-theory doesn't have to be tiny at all. It can be very big, though undetectable because it lies in a direction at right angles to all the familiar three dimensions of space. In the same way that a genuinely two-dimensional creature inhabiting a two-dimensional world would be unaware of the third dimension,* we three-dimensional creatures would be unaware of this fourth (or tenth!) dimension of space.

The way this changes our mental image of what is going on is that instead of thinking of particles as being the detectable consequences of vibrating strings we have to think in terms of vibrating sheets, or membranes. For this reason, although Witten has never specified what the "M" in "M-theory" stands for, to many people it stands for "membrane." More technically, a sheet in two dimensions is called a two-brane, and in these models there are equivalent structures (though harder to visualize) for all dimensions up to ten, generically known as p-branes, where p can be any whole number less than ten. A string would be a "one-brane."

One consequence of all this is that our entire Universe could be a three-brane embedded in higher dimensions. And that raises the possibility that there might be other three-dimensional universes, right alongside us, also embedded in higher dimensions but completely inaccessible to us, rather in the way that you could regard the surfaces of the pages of a book as a series of two-dimensional universes, right next door to each other but each seeming, to any two-dimensional creatures that inhabited them, to be the entire world.

---

*A situation delightfully described in Edwin Abbott's classic *Flatland* (1884) and in Ian Stewart's modern reworking of the tale.

Such ideas take us into the realm of speculation, albeit respectable speculation, with no immediate prospect of testing them in any experiment or accelerator we could build on Earth. But we do have access to information from an event in which the conditions were so extreme that the influence of M-theory processes, strings, and membranes may have left a mark. Our best understanding of the Universe in which we live suggests that it emerged from a state of extremely high pressure and temperature, the Big Bang, about 14 billion years ago. Astronomical observations are now so accurate that scientists can use these data to test some of the predictions from particle theory about what happened in the Big Bang itself. Cosmology and particle physics have merged to become astroparticle physics. So the logical next step in probing the behavior of matter on the smallest scales is to look outward into space at the behavior of matter on the very largest scale, the scale of the Universe itself, and to investigate where it all came from.

3

How Did the Universe Begin?

It is now widely accepted that the Universe we inhabit emerged from a hot, dense fireball called the Big Bang. In the 1920s and 1930s, astronomers first discovered that our Galaxy was simply one island of stars scattered among many similar galaxies and then that groups of these galaxies were moving apart from one another as the space between them stretched. This idea of an expanding universe had actually been predicted by Einstein's general theory of relativity, completed in 1916, but it had not been taken seriously until the observers made their discoveries. When it was taken seriously, mathematicians discovered that the equations exactly describe the kind of expansion we observe, with the implication that if galaxies are getting farther apart as time passes then they must have been closer together in the past; and long ago all the matter in the Universe must have been piled up in a dense fireball.

It is the combination of theory and observation that makes the idea of the Big Bang so compelling; clinching evidence in support of the idea came in the 1960s with the discovery of a weak hiss of radio noise, the cosmic background radiation, that comes from all directions in space and is interpreted as the leftover radiation from the Big Bang itself. Like the expansion of the Universe, the existence of this background radiation was predicted by theory before it was observed experimentally. By the end of the twentieth century, the combination of theory and observations had established that the time that has elapsed since the Big Bang is about 14 billion years, and that there are hundreds of millions of galaxies like our own scattered across the expanding Universe. The question cosmologists are now confronting is how the Big Bang itself occurred — or, if you like, how did the Universe begin?

The starting point for confronting this question is the cosmologists' own standard model, which combines everything they have learned about the expanding Universe from observations with the theoretical understanding of space and time incorporated in Einstein's general theory. Establishing this model has been helped by the fact that the farther out into the Universe we look, the farther back in time we see. Because light travels at a finite speed, when we look at galaxies millions of light years away, we see them as they were millions of years ago, when the light now reaching our telescopes left them. With powerful telescopes, astronomers can see what the Universe looked like when it was younger — and the cosmic background radiation allows us to "see" (with radio telescopes) the last stage of the fireball itself.

If we imagine winding the expansion of the Universe backward, it seems that there must have been a time when everything was piled up in a single point of infinite density, called a singularity. This naive image of the birth of the Universe is borne out by the general theory of relativity, which says that the Universe must indeed have been "born" in a singularity. But, as we have mentioned, physicists are uncomfortable with the idea of singularities and infinities and usually regard any theory that predicts their existence in the physical Universe as flawed. This is true even of the general theory. It can tell us how the Universe as we know it emerged from a state of *nearly* infinite density; but it cannot tell us what actually happened at the very beginning, the moment of the Big Bang itself. (I use *Big Bang* to refer to this moment of birth as well, although strictly speaking the term refers to the hot fireball phase that followed a few moments later.) The standard model of cosmology can tell us that this moment occurred about 14 billion years ago, and it can describe everything that happened after that moment. We can set the moment where the general theory breaks down as time zero and count forward from it to describe the evolution of the Universe.

The farthest back we have yet seen, to the origin of the background radiation, corresponds to a time a few hundred thousand years after the moment of the Big Bang, when the entire Universe was filled with hot gas (technically known as plasma) that was about the same temperature as the surface of the Sun today, a few thousand degrees Celsius. At that time what is now the entire visible Universe was just one-thousandth of its present size, and there were no individual objects on the scale of stars or galaxies in the maelstrom of hot material. But tiny differences in the temperature of the background radiation can be observed at different places on the sky today, and these irregularities tell us about the size and nature of the irregularities that existed in the Universe at the end of fireball phase. Moving forward from that time, the irregularities observed in the background radiation are just the right size and pattern to explain the origin of galaxies and groups of galaxies — they are the seeds from which the structure we see in the Universe today grew. (More of this in later chapters.) Moving backward from that time, the pattern of irregularities we can see in the background radiation tells us about the kinds of irregularities that were present when the Universe was even younger, at the time when the general theory itself breaks down.

The first and most important thing about these irregularities in the background radiation is that they are tiny. They are so small that at first it was impossible to measure them, and the radiation seemed to be coming perfectly uniformly from all directions in space. If the radiation had been *perfectly* uniform, the whole standard model would have fallen apart, since if there had been no irregularities in the Big Bang fireball there would have been no seeds from which galaxies could grow, and we would not be here. The fact that people are around to puzzle over such questions told astronomers that there must be irregularities in the background radiation, if they could only develop instruments sensitive enough to measure them; but it was not until the early 1990s, nearly thirty years after the background radiation had been discovered, that NASA's COBE satellite was able to make the sensitive measurements needed to show that there are indeed tiny ripples in the background radiation. The two key questions posed by this discovery are, Why is the background radiation very nearly smooth? and What made the ripples?

The first question is more profound than you might think, because even today, 14 billion years after the Big Bang, the Universe is very nearly smooth. This isn't obvious if you contrast the brightness of a galaxy like our Milky Way with the darkness of space between the galaxies, but it soon becomes apparent on larger scales. The Universe isn't *exactly* uniform, but even in terms of the distribution of galaxies it is uniform in the same way that a perfectly baked raisin loaf is uniform — no two slices of bread have exactly the same pattern of raisins in them, but every slice looks much the same as every other slice. In the same way, if you take a photograph of the galaxies that can be seen in a small patch of sky, it will look much the same as a photograph of the same-sized patch of a different part of the sky. The background radiation is even more uniform, and looks exactly the same from all parts of the sky, to within a fraction of 1 percent. The profundity of this observation lies in the fact that there has not been enough time since the Big Bang for all the different parts of the Universe to interact with one another and become smoothed out.

Taking the most extreme example, the background radiation from one side of the sky has been traveling for 14 billion years to reach us, and the radiation from the other side of the sky has also been traveling for 14 billion years to reach us, but both lots of radiation have almost exactly the same temperature. Since this radiation (electromagnetic energy) travels at the speed of light, and nothing can travel faster than light, there seems to be no way in which the opposite sides of the sky could "know" what temperature they ought to be in order to ensure this uniformity. Some great conspiracy seems to have been at work to make the cosmic fireball the same everywhere, even though different parts of the fireball could not interact with one another.

This homogeneity is related to another puzzling feature of the Universe: its flatness. The general theory of relativity tells us that space (strictly speaking, space-time) can be bent and distorted by the presence of matter. Locally, near an object like the Sun or the Earth, this distortion of space-time produces the effect we call gravity. Cosmically, in the space between the stars and galaxies the combined effect of all the matter in the Universe can produce a gradual curve in space in one of two ways.

If the density of the Universe is more than a certain amount (called the critical density) then three-dimensional space must curve around on itself the way the two-dimensional surface of a sphere is curved, to make a closed surface. It doesn't matter how much the density exceeds the critical density, just that it does exceed it. Such a space is finite but unbounded, like the surface of the Earth. The Earth's surface has a finite area, but you can keep going along it in any direction for as long as you like without ever coming to an edge — you just keep going round and round the globe. If the Universe is like that, it must have a finite volume, but if you keep moving in any direction you will never get to an edge, although you will (eventually) get back to where you started.

The other possibility is that the density is less than the critical density. It doesn't matter how much less, just that it is below the critical value. In that case, the Universe is "open," with space curved outward, in a way like the shape of a saddle or a mountain pass, but going on forever. Such a universe would be infinitely big, and you could keep going in a straight line forever without ever visiting the same spot twice.

Exactly in between these two possibilities there is a unique case, a so-called flat universe. This would have exactly the critical density, and space would be a three-dimensional equivalent of a flat surface.

The three possibilities correspond to three different possible fates for the Universe. In a closed universe, the gravitational influence of all the stuff in the universe will gradually halt its expansion, and make it collapse back into a fireball reminiscent of the Big Bang (sometimes called the Big Crunch). If the Universe is open, it will expand forever and never stop. But if it has exactly the critical density, it will expand at a slower and slower rate (with one exception, which we shall come to later) until, in the far distant future, it hovers in a state of suspension, neither expanding nor collapsing, balanced on a gravitational knife-edge.

By the mid-1970s, it was clear from observations of the expanding Universe that if the critical density were defined as 1, then the density of the real Universe lay somewhere between 0.1 and 1.5, very close to the *only* special density allowed by the general theory. This was puzzling enough, since at that time there was no reason to think that the Universe could not have emerged from the Big Bang with any density at all; but to make the puzzle more profound cosmologists then realized that the expansion of the Universe always forces it *away* from the critical density as time passes. A closed universe becomes "more closed" and an open universe becomes "more open" as it expands. The fact that the density observed today is so close to 1 means that just one second after the Big Bang the density must have been within one part in 10^{15} (that is, one part in a million billion) of 1. It was between 0.99999999999999 and 1.00000000000001. The only explanation seemed to be that something must have ensured that the Universe started out with *exactly* the critical density, and that the density is still exactly 1, in these units today. But what could have forced the Universe toward such uniformity and flatness at the moment of its birth?

Using the equations of the general theory of relativity, we can work backward in time from the fireball stage of the Universe to calculate the temperature and density of the Universe at earlier times. This tells us that one-ten-thousandth of a second (10^{-4} seconds) after the moment of the Big Bang the entire Universe was at the density of the nucleus of an atom today (10^{14} grams per cubic centimeter) and had a temperature of 10^{12} K (1,000 billion degrees).* Atomic nuclei have been studied for a hundred years, and as we saw in Chapter 2, conditions like these have been studied in particle accelerators for decades.

---

*Temperatures on the Kelvin scale are measured in degrees that are the same size as those on the familiar Celsius scale but start from the absolute zero of temperature, -273 °C. The abbreviation is K, not °K. So 273 K is equivalent to 0 °C, and so on.

Physicists are sure that they understand what happens to everyday matter under such conditions and at all the less extreme conditions that developed as the Universe expanded and cooled from that state. So we are confident that we know about the evolution of everyday matter from 10^{-4} seconds after the moment of the Big Bang onward.

Some of the details of this will be discussed later; what matters now is that both the extreme smoothness and the flatness of the Universe, and the tiny irregularities that grew to become clusters of galaxies today, must have already been imprinted on the Universe at that time, since there is no mechanism that could have imposed them later. This means that the calculations have to be extended farther back in time, into conditions of temperature (energy) and density that have been probed by some accelerator experiments, but which are not as well understood as nuclear matter is. We *think* we know what happened at even earlier times, but this is still work in progress. The closer we extend those calculations to the moment of the Big Bang, the more speculative they become.

Eventually, the general theory of relativity breaks down and cannot be used anymore. This happens at the point where quantum physics dominates, and the idea of a smoothly continuous space-time (the stretched rubber sheet) that is at the heart of the general theory itself breaks down. According to quantum theory, space and time themselves are quantized, and it is meaningless to talk about any distance shorter than 10^{-35} meters (a distance dubbed the "Planck length") or any time shorter than 10^{-43} seconds (the "Planck time"). So there was no singularity (with zero length, at time zero), and we should picture the entire observable Universe as having been "born" 10^{-35} meters across, with a density of 10^{94} grams per cubic centimeter, and an "age" of 10^{-43} seconds. It is meaningless, in this context, to talk of earlier times, shorter lengths, or greater densities.

What we think happened next draws heavily on the ideas of grand unification. The theories we described in Chapter 2 predict that when the Universe was born all the forces of nature were on an equal footing, but they soon split apart from one another. As they did so they gave the Universe a violent outward shove which smoothed and flattened it in a process called inflation while leaving exactly the right kind of ripples to explain the irregularities in the background radiation and the presence of clusters of galaxies today.

The process in which the forces of nature split apart from one another as the Universe cooled is analogous to a phase transition like the one when water freezes into ice. During such a phase transition, energy is exchanged between the system that is changing and the world at large. For example, when ice is at 0 °C, and it is melting, it stays at 0 °C all the time it is melting, even though it is surrounded by warmer material and is absorbing energy. All the energy it is absorbing goes into melting the ice, not into warming it up. When water freezes, this process is reversed. Water at 0 °C stays at 0 °C while it is freezing, even though the outside world is colder. Heat — known as latent heat — is released by the water as it freezes, and it is this same latent heat which has to be replaced if you want to melt the ice. There is an even greater release of latent heat when water vapor condenses into liquid water, and the heat released through this process when raindrops form drives the convection which builds thunderclouds. In the very early Universe, a kind of superconvection occurred. Gravity split off from the other forces at the Planck time, 10^{-43} seconds, and had split from the strong nuclear force by about 10^{-35} seconds. Together, these phase transitions released a huge amount of energy that made the Universe expand exponentially for a fraction of a second. But a fraction of a second was all it took for inflation to do its work.

There's a neat analogy for this process. Imagine a lake high in a glaciated mountain region, walled off behind a dam of ice. The lake may be full of deep, calm water in equilibrium, not going anywhere. In scientific terms, the water is said to occupy a local minimum in terms of the energy associated with the gravitational field of the Earth. But this calm stability disguises the fact that the water in the lake stores a great deal of gravitational potential energy, so that in a sense this is a false minimum; if the water could get out of the lake it would rapidly flow downhill to the sea, which represents a true minimum (at least, as far as the surface of the Earth is concerned). Now imagine a climate change, or even just the seasonal change from winter to summer. The ice dam melts, and the water in the lake is released to pour down the mountainside in a torrent, eventually reaching the sea and settling into equilibrium once again, but at a lower energy level. Physicists describe the conditions in the Universe before inflation as a false state of equilibrium in terms of energy associated with the vacuum (the energy of empty space, or space-time, if you like). This "vacuum energy" is released during the phase transition that drives inflation, as the Universe settles down into the true minimum energy state of the vacuum. Inflation itself is like the torrent of water rushing from one energy level to the other, a short-lived episode between two different equilibrium states.

In the case of inflation, *very* short-lived. Inflation lasted for only about 10^{-32} of a second, but during that time the size of what would become the visible Universe doubled once every 10^{-34} seconds (some versions of the theory suggest even faster growth, but this is ample for our needs). In other words, in that 10^{-32} of a second there were at least a hundred doublings (10^2, because $34 - 32$ is 2). This was enough to take a volume just 10^{-20} times the size of a proton and inflate it to a sphere about 10 centimeters across, roughly the size of a grapefruit. This is equivalent to taking an object the size of a tennis ball and inflating it to the size of the present-day observable Universe in that same time, 10^{-32} of a second. As that comparison makes clear, one of the peculiarities of this inflation is that in a sense it proceeds faster than light. Even light would take 3×10^{-10} of a second to cross a space 1 centimeter across, but inflation expands the Universe from a size much smaller than a proton to a sphere 10 centimeters across in about 10^{-32} seconds. This is possible because it is space itself that is expanding — nothing is traveling "through space" at this speed. And this is why the Universe is so uniform. Everything we can see came from a seed of energy so small that there was no room inside it for any significant irregularities to exist, and inflation froze this original uniformity into the cosmic grapefruit that settled down into a more steady expansion, coasting on the momentum left over from inflation, as the energy from the phase transitions was dissipated.

Inflation also explains why the Universe is so flat. Stretching things (even space-time) tends to smooth out wrinkles and curves. Think of a wrinkly prune swelling up as it soaks in water to become a smooth sphere. Or imagine the Earth, which already seems pretty flat to anyone living on it, expanded to the size of the Solar System. It would be very hard to tell that you were living on the surface of a sphere if the planet were that big. Whether the seed of the Universe was closed or open, inflation would drive it so close to flatness that after a hundred doublings or more it would be impossible for any instruments we could devise today to measure the tiny deviations from flatness.

This was both a triumph of inflation theory and, when it was first proposed in the early 1980s, something of an embarrassment. It seemed like too much of a good thing. At that time, for reasons we shall go into later, astronomers thought that the density of the Universe was about one-tenth of the critical density. But inflation predicted that the density of the Universe should be indistinguishably close to the critical density required for flatness. Either there was something wrong with the idea of inflation, or there must be a lot more stuff in the Universe than the astronomers of the early 1980s had accounted for. The natural reaction at first was that inflation, the new kid on the block, must be wrong. But increasingly sophisticated studies of the background radiation, culminating in the observations made by NASA's WMAP satellite early in the twenty-first century and ESA's Planck Explorer a little later, showed that the Universe really is indistinguishably close to flatness, so its density must be indistinguishably close to the critical density. This left the puzzle of where the "missing" mass (sometimes called dark matter, since it has never been seen) was; the resolution of that puzzle forms the subject of Chapter 6.

Inflation theory is still a work in progress, and as with GUTs there are different variations on the theme. But its overall successes, and in particular its successful prediction that when accurate enough instruments were developed the Universe would be found to be precisely flat, tell us that there is something fundamentally true about the inflation concept, even if we don't know which version of the theory (if any of the current ones) will eventually emerge triumphant. Among its other successes, the theory also tells us where the tiny irregularities that grew to become clusters of galaxies came from, and in the same breath hints at a possible mechanism for the origin of the Universe itself. This mechanism has to do with the quantum fluctuations of the vacuum that we encountered earlier.

Quantum uncertainty means that on the smallest scales it is not possible for the Universe to be perfectly smooth and regular. There must always be tiny irregularities, roughly on the scale of the Planck length, popping into existence and disappearing again. Such quantum fluctuations have little effect on our everyday world today, at least on human length scales (although they may be important in under-standing the nature of the force that operates between electrically charged particles such as electrons and protons, so in that sense they are definitely relevant to our everyday lives). But cosmologists real-ized that these fluctuations must have been going on at the time of inflation itself. The fluctuations from a time when what is now the entire visible Universe was roughly a hundred million times bigger than the Planck length would have been stretched by inflation, before they could flicker out of existence, to form a network of irregularities that filled the Universe when it was the size of a grapefruit, at the end of inflation. These irregularities would be imprinted on the Universe and remain throughout the fireball phase, stretching with the univer-sal expansion, right up to the time, a few hundred thousand years after the Big Bang, when the Universe had cooled to the temperature of the surface of the Sun today and the cosmic background radiation was set on its way across the Universe. Quantum theory makes precise predictions about the kind of pattern of irregularities that would be produced by this process, and, statistically speaking, this is exactly the same as both the pattern of irregularities seen in the background radiation itself and the pattern of irregularities in the distribution of galaxies across the sky seen on the largest scales. This is another re-markable triumph of inflation theory — it predicts that the Universe will be very nearly perfectly smooth, but that it will contain the kinds of irregularities required for galaxies to grow as the Universe expands. And it means that the largest irregularities in the Universe (superclus-ters of galaxies) have their origin in the smallest possible irregularities that can exist, quantum fluctuations of the vacuum.

Indeed, the entire Universe could have grown from a quantum fluctuation of the vacuum, thanks to a combination of inflation and a curious property of gravity.

This curious property of gravity is that it stores *negative* energy. When something (anything!) falls downward in a gravitational field (like the water rushing out from the mountain lake described earlier) energy is released. That energy comes from the gravitational field. When the object (in this case, the water) is higher up it has more potential energy than when it is lower down. The difference between the two energy levels explains where the energy comes from that makes the water move. But where do you measure the energy levels? The gravitational force between two objects is proportional to 1 over the square of the distance between them. So this force is zero when they are infinitely far apart, since 1 divided by infinity (let alone by infinity squared) is zero. In the Einsteinian picture, this is equivalent to saying that the gravitational influence of an object disappears at infinity because space-time at infinity is completely undistorted by the mass of the object. Either way, it means that the energy associated with an object in a gravitational field is zero when it is infinitely far away from the source of the field. But we have already seen that when an object moves downward in a gravitational field (that is, when it gets closer to the source of the field) it takes energy from the field and turns it into energy of motion (the water rushing down the mountainside, or a cup dropped from your hand, or anything falling down under the influence of gravity). That energy comes from the gravitational field itself. The field starts with zero energy, and gives energy to the falling object, so that the field itself must be left with negative energy. This is literally true — it is not some trick of the equations, since we don't have any choice about where we measure the zero point energy of the field from. But what has this got to do with quantum fluctuations?

There is no limit, in principle, to how much mass (strictly speaking mass-energy, bearing in mind $E = mc^2$) a quantum fluctuation can have, although the more massive (energetic) a fluctuation is, the less likely it is to occur. In the early 1970s, the American cosmologist Ed Tryon pointed out that in principle a quantum fluctuation containing the mass-energy of the entire visible Universe could arise out of nothing, and that although the mass-energy of such a fluctuation would be enormous, in the right circumstances the negative gravitational energy of the gravitational field associated with all that mass would exactly balance this, so that the overall energy of the fluctuation was zero.

At the time, this seemed like a meaningless mathematical trick, since it was "obvious" that a tiny quantum entity with such a powerful gravitational field could never expand and would snuff itself out of existence as soon as it appeared. Ten years later, however, inflation, which in many ways is a kind of antigravity effect, offered a way in which a quantum fluctuation containing enough energy to make all the matter in the Universe could have been inflated to the size of a grapefruit and left with a residual outward expansion before gravity had time to snuff it out of existence. In an expression made popular by the pioneering inflation theorist Alan Guth, a universe that could appear out of nothing would be "the ultimate free lunch." It happens that the right circumstances for gravity to balance matter in energy terms in this way include the requirement that the universe be closed, in the sense we described earlier, although it is allowed to be indistinguishably close to being flat. All of this accords with observations of the Universe we live in.

By now, we have moved into the realms of scientific speculation, although it is still respectable speculation. But it is impossible to stop here, since such ideas beg the question of where, if the Universe did begin in this way, the original quantum fluctuation itself came from. Today there are almost as many suggestions in answer to that question as there are cosmologists, and any of them — or none of them — may be correct. But in the spirit of providing some insights about things we still *don't* know, here is my personal favorite such speculation, and the one getting the most attention from the experts just now.

My personal favorite is the idea (which itself comes in many different forms) that the kind of quantum fluctuation that gave birth to our Universe could happen anywhere in our Universe today. This would not mean a violently expanding fireball the size of a grapefruit blasting outward somewhere in our space-time, because although the process could start in our Universe (perhaps triggered by the collapse of a massive star into a black hole) it would then expand into its own set of dimensions, all of them at right angles to all of the dimensions of our Universe. The implication, of course, is that our Universe was born (or budded off) in this way from the space-time of another universe, and that there was no beginning and will be no end, just an infinite sea of interconnected bubble universes. It is even possible that before too long — perhaps in a hundred years or so — we might have the technological ability to create universes in this way, and that our Universe may have been deliberately created by intelligent beings in another universe, as an experiment of some kind. But that is far enough to take such speculations here.*

The hottest idea being discussed at present, in the first decade of the twenty-first century, to account for the birth of the Universe kicks off from the ideas of strings and membranes discussed in Chapter 2. One particular variation on that theme envisages our Universe as a ten-dimensional entity in which all but three dimensions of space and one dimension of time are compactified, rolled up so small that we cannot

---

*For more about these ideas, see my *In the Beginning* and Lee Smolin's *The Life of the Cosmos*.

detect them directly. This entire Universe could be a membrane which floats about in an eleventh dimension, in a manner analogous to the way in which a two-dimensional surface of a sheet of paper can be moved around in the third dimension of space. There could be many of these membrane universes sharing the same eleventh dimension, just as there can be many pages with two-dimensional surfaces in a fat three-dimensional book. Like the pages in a book, these universes could be very close to one another. When two sheets of paper are laid one on top of the other, every point on one sheet is next to a point on the other sheet, and in the same way every point of space in our three-dimensional Universe may be next door to a point in another three-dimensional universe just a tiny bit away from us in the eleventh dimension. The universe next door might be only a fraction of a millimeter away from us, as close as your underclothes are to your skin, but in a direction that makes it impossible for us to see it or communicate with it. Actually, in a sense, the universe next door is even closer than that, because it doesn't just surround you like a second skin—every point in *three-dimensional* space, including all the ones *inside* your body, lies next to a point in the other universe.

One way to get a handle on this is to go back to the idea of a universe as a flat, two-dimensional sheet, but now inhabited by two-dimensional creatures in the shapes of geometrical figures like squares, triangles, and so on. Such beings are literally two-dimensional, and have no way of perceiving anything going on in the third dimension, off the paper. The inside of a square creature would be its equivalent of the inside of your body in three dimensions, and if a three-dimensional being were cruel enough to poke the inside of the square the creature would feel an internal pain apparently without a cause. If a three-dimensional sphere approached this two-dimensional world and slowly passed through it, the inhabitants of flatland would first see a point as the pole of the sphere touched the plane they live in. As the sphere moved onward, this point would become a circle that grew until the equator of the sphere reached the plane, then shrank away again, finally disappearing at a point as the other pole of the sphere completed its passage through the plane.

But what would happen if another entire flat universe approached the flatland and touched it? Depending on just how you set up the equations, the two universes could pass through one another without any effect, or they could interact violently. When we extend this image to three-dimensional universes (plus one time dimension and several compactified dimensions) moving in the eleventh dimension, and we apply the various constraints on the equations that we have learned from the search for a theory of everything, we find that when two empty and rather dull universes collide in this way the event can trigger just the kind of quantum outburst that, with the aid of inflation, turns into a universe like our own. The situation can be more complicated (and therefore more interesting) because the two universes need not be exactly flat to start with — imagine two sheets of paper that have been crumpled then roughly smoothed out approaching one another, and you will see that different points on the two surfaces touch before most of the two worlds are in contact. And, of course, the space-times of the two universes need not be flat in the way we have described at all; they could be curved, like the surface of a sphere, or a toroidal doughnut shape, or shaped in other interesting ways. All of this gives scope for plenty of speculation among M-theorists about how the Universe we live in might have begun. Until (if ever) there is some reason to think that one variation on the theme has some bearing on reality there is not a lot of point in elaborating on them outside such specialist circles. But one of the simplest ideas certainly provides food for thought, and shows that there may be some point in discussing what happened before the Big Bang.

In many versions of M-theory, gravity, alone of the four forces of nature, extends out of our Universe and into the eleventh dimension. In physical terms, these ideas are appealing because they could explain why gravity is so much weaker than the other forces — it is because, in a sense, most of the effect of gravity leaks away from our three-dimensional world. You can make an imprecise, but helpful, analogy with a two-dimensional metal plate that is suspended in a tank of water. If you hit the plate with a hammer, sound waves will reverberate through the plate, but some of the energy in the waves will leak away in the form of sound waves traveling through the third dimension, into the water. So there will be less energy in the plate itself.

According to some M-theory models (which go by the name "ek-pyrotic universe," and come back into our story later), if two empty three-dimensional universes floating in the eleventh dimension approached one another, they would be pulled together by gravity and collide. This would trigger events like our Big Bang in both universes, but the release of energy would also make the universes bounce off each other and set them drifting apart in the eleventh dimension. While they drifted in the eleventh dimension, each universe would expand in its own three dimensions, with matter being spread ever more thinly until it resembled the state the universes were in before the collision. Eventually, though, gravity would overcome this drift and pull them back together, triggering another set of big bangs and bounces, and so on indefinitely. This idea is discussed further in Chapter 10.

Like many of the ideas discussed by cosmologists trying to understand where the Big Bang came from, these ideas imply that our

Universe is not unique, and that our Big Bang was not unique. But it may, in a sense, be special. There may be an infinite number of other universes in some sense alongside ours, and there may have been, or may yet be, an infinite number of big bangs extending both backward and forward in time. But they are unlikely to be all the same. Some universes may expand only a little way out of their big bang before collapsing again; some may expand so rapidly that matter is spread too thin ever to form galaxies, stars, and people. It is quite possible that in other universes the forces of nature have different strengths from those in our Universe, so that nuclear reactions proceed more slowly, or more rapidly, and one way or another the kind of complex molecules that make up our bodies never get a chance to form.

People have long been puzzled by the fact that our Universe is in many ways just right for the emergence and evolution of life. Some have argued that the Universe has been designed for life, an idea which gains some credence from the possibility that it was created as a lab experiment in some other universe (but then, how come that universe was just right to allow the emergence of intelligent life?). Others, though, have suggested that in an infinite number of universes every possible combination of laws of physics and forces of nature must exist somewhere, or somewhen. Most of this infinite array of universes would be sterile, because the detailed conditions required for life do not exist in them. But some, by chance, will indeed be just right for life, in the same way that baby bear's porridge was just right for Goldilocks, even though it was not made for her. Life forms similar to ours will exist only in universes where conditions are right for life, so it is no surprise that we find our Universe so suitable for ourselves. It is also part of the nature of infinity that with an infinite number of worlds to choose from, although our kind of world might be very rare, even a small fraction of infinity is itself infinity, which makes us special, but not all that special. If these ideas are correct, there must be an infinite number of these Goldilocks universes where life forms similar to ours exist. It's like the difference between having a suit tailor-made for you, and unique, or choosing one off the peg. If there is an infinite variety of suits in all possible shapes and sizes to choose from, there is no point in having a suit made since there must be one that is a perfect fit ready and waiting for you, like that porridge.

In either picture, though, the fact is that we live in a Universe where there are certain rather well-understood laws of physics, and four forces of nature with very well-studied properties. Leaving debate about what went on before the Big Bang aside, we know what the Universe was like a fraction of a second after the moment it was born, at the end of inflation, when it was the size of a grapefruit, streaked with blown-up quantum irregularities, very hot and still expanding rapidly but with gravity just beginning to slow the expansion down. The next question is, how did the early Universe develop out of that fireball?

4

How Did the Early Universe Develop?

The same processes that drove inflation may have been responsible for the production of the matter that makes up stars, planets, and ourselves in the Universe today. Most of the mass of this everyday material is in the form of protons and neutrons (collectively known as baryons), which are themselves composed of quarks. The other important component of everyday matter is the lepton family, dominated today by electrons and neutrinos. Because of the overwhelming contribution of baryons to the mass of the visible Universe, however, everyday matter is often referred to simply as baryonic matter. The seed from which our Universe developed was an immensely hot, immensely dense fireball of pure energy. The question is, how did this fireball give rise to the kind of baryonic matter we see all around us as the Universe expanded and cooled? Or, if you like, where do quarks and leptons come from? We think we know the answer, although, as ever, the explanation is more speculative the farther back in time we look and the higher the energies that have to be considered.

The various degrees of speculation that are involved can be gauged by comparing the energy density of the early Universe at different times, calculated by winding back the observed expansion of the Universe (like winding the hands of a clock backward to earlier times) in accordance with the equations of the general theory of relativity, and looking at conditions with the energy densities (or average energy per particle) that have been achieved in different generations of particle accelerators over the years. These energies are measured, as usual, in electron volts (eV), and a useful benchmark to bear in mind is that the mass of a proton is just under 1 GeV (a thousand million eV) which corresponds to 1.7×10^{-27} kilos. We can also make a comparison between the density of the Universe at different epochs and the density of water, which is 1 gram per cubic centimeter.

A convenient place to end the story told in this chapter will be at the time, a few hundred thousand years after the Big Bang, when the Universe cooled to below the temperature of the surface of the Sun today (roughly 6,000 K, or a mere half an electron volt) and the radiation detected today as the cosmic microwave background began to stream freely through space. At that time, the density of the Universe was only 10^{-19} (one-tenth of a billionth of a billionth) of the density of water, and there is no doubt that we understand the behavior of matter under such conditions. Of course, there were no "days," "weeks," or "years" at that time, long before the Earth or any other planet existed, but treating these units simply as measures of time, each corresponding to a certain number of seconds, we can say with confidence that a year after the Big Bang the temperature of the Universe was two million K, even though its density was still a little less than a billionth of the density of water. A week after the Big Bang, the entire Universe was at a temperature of 17 million K, about 10 percent hotter than the center of the Sun today, and although the density was only a millionth of the density of water, the pressure in the fireball was more than a billion times the pressure of the atmosphere at the surface of the Earth today.

The next landmark brings us to conditions similar to those probed by the first cyclotron, built in the early 1930s. Two hundred seconds (a little over three minutes) after the Big Bang, the average energy per particle in the Universe was 80,000 eV (80 keV), equivalent to a temperature of just under a billion K. Since we have been carrying out experiments with at least this much energy per particle for more than seventy years, we are confident that we understand the particle inter-actions that were going on at that time — and, indeed, much earlier times. One second after the Big Bang, the temperature of the Uni-verse was about ten billion K (nearly a million eV), and the condi-tions everywhere resembled the conditions at the heart of an explod-ing star, a supernova, today. The density of matter was 500,000 times the density of water, and the pressure was 10^{21} times the atmospheric pressure on Earth today. The last landmark that keeps us in touch with everyday matter in its present form lies at a time 10^{-4} seconds (one-ten-thousandth of a second) after the start of the Big Bang (time zero), when the density of the Universe was roughly the same as the density of an atomic nucleus today, and the temperature was about a thousand billion K (10^{12} K, or about 90 million electron Volts, 90 MeV). Such conditions are so well understood, and have been for so long, that the story of the Universe from that time for-ward, the standard Big Bang model, was well established by the end of the 1960s.

Even before then, particle accelerators operating in the 1950s and

1960s had reached energies of several GeV, corresponding to temperatures (as far as the concept has any meaning at such high energies) in excess of 30×10^{12} K. (For comparison, the temperature at the heart of a star corresponds to less than one-tenth of 1 GeV.) Such conditions existed in the Universe about 3×10^{-8} seconds (30 billionths of a second) after time zero. In the 1980s, an accelerator called the Tevatron, at Fermilab, near Chicago, reached an energy of 1,000 GeV, reproducing for a split second conditions that existed in the Universe when it was just 2×10^{-13} seconds old. Such accelerators provided the experimental input used in the development of the theories of particle physics described in Chapter 1. The theorists have been able to speculate even further with their ideas about grand unification, supersymmetry, and membranes, which may provide insight into what happened before the Universe was 10^{-13} seconds old; the next direct step in testing those theories here on Earth is taking place near Geneva, where CERN's Large Hadron Collider is just beginning to operate. It will, if all goes well, reach energies in excess of 7,000 GeV, probing conditions that existed in the Universe 10^{-15} seconds after time zero. It is still a big and speculative leap back from that time to 10^{-39} seconds; but we *think* we know, at least in outline, what was going on even then. And that is where we have to pick up the story and move forward in time if we are to understand where baryonic matter came from.

The story starts with the grand unified theories and their X bosons, which are involved, if the models are correct, in proton decay. At 10^{-39} seconds after time zero, the average energy per particle was about 10^{16} GeV and the temperature was 10^{29} K. The density was 10^{84} times the density of water, which corresponds to packing 10^{12} (a thousand billion) stars like the Sun into the volume of a single proton. It is under those conditions that the X bosons come into their own.

When we mentioned the creation of virtual particles out of energy earlier, we skimmed over one important feature of the process. There are some properties of particles, such as electric charge, that seem to be conserved in the Universe. Overall, electric charge cannot be created or destroyed in any experiments, nor in any natural process observed in the Universe today, so there is always exactly the same amount of charge, as far as we can tell, in the world (as it happens, zero). If you want to make a negatively charged particle, such as an electron, out of energy, you also have to make a positively charged particle to balance the books. In this case, the positively charged counterpart to an electron is called a positron, the anti-electron that we met earlier, and it has the same mass as an electron but carries one unit of positive charge. So when we talked of a cloud of virtual particles surrounding a charge, we were referring more specifically to a group of electron-positron pairs swarming around the charge, not just electrons.

It isn't just the electric charge that is different for particles and antiparticles, though. There are other quantum properties of particles which are conserved in the same way, and as we hinted in Chapter 1 each variety of particle in the world today is thought to have an "antiparticle" counterpart with all the opposite properties. This does not mean that every single particle has an antiparticle partner, but that in principle it is possible for such an antiparticle to exist if the energy is available to make the requisite particle-antiparticle pair. A sufficiently energetic photon (one that contains more than the rest-mass energy of two electrons) can turn itself into a *pair* of particles, an electron and a positron. But when a positron meets an electron, both particles disappear in a puff of high energy photons — gamma rays — as their opposite quantum properties cancel each other out.

The equivalent processes affect all particles. They can be made in matter-antimatter pairs out of pure energy, but they annihilate each other and release that energy again when particle and antiparticle counterparts meet. The particles don't even have to be electrically charged — there is an antimatter counterpart to the neutron, for example — but charge is the most obvious and convenient label with which to distinguish antimatter entities from their matter equivalents if they do happen to be charged. (Of course, neutrons are made of quarks, and antineutrons are made of antiquarks; but there are genuinely neutral quantum particles which also have antimatter counterparts.)

When the Universe was born, it was in the form of pure energy. But that energy immediately began to make particle-antiparticle pairs, and the particle-antiparticle pairs immediately began to annihilate one another to make energy once again. As the Universe expanded and cooled, the amount of energy available in each tiny volume of the Universe fell, and as the energy density fell it became impossible to make the more massive particles. Eventually, the energy density (equivalent to temperature) would have fallen to the point where it would have been impossible even to make electrons. If the processes we have described so far were perfectly reversible, as they seem to be in almost all experiments carried out on Earth, the result would have been a young universe that contained equal numbers of matter and antimatter particles, when it was still so dense that collisions between particles would have been frequent. For every electron there would have been a positron, for every quark an antiquark, and so on. Each particle would have met up with an antiparticle counterpart and been annihilated. By the time such a universe was a few hundred thousand years old, all the matter would have turned itself back into radiation, but it would now be at a temperature too low for any more particle pairs to be created. There would be no matter in the universe at all. So where did the stuff we are made of, and the stuff that makes up all the stars and galaxies in the visible Universe, come from?

The only possible answer is that under the conditions that existed early in the life of the Universe, the processes we have just described were not completely symmetric. The first person to appreciate this fully, and to spell out its implications in simple terms, was the Russian physicist Andrei Sakharov in the 1960s.

The jumping-off point for Sakharov's suggestion was an experimental discovery made in the early 1960s that took the world of particle physics completely by surprise. It had to do with a property of quantum particles called, for historical reasons we need not go into here, "CP symmetry." The simplest way to get a picture of what CP symmetry is all about is to imagine an interaction involving quantum particles, then imagine replacing every particle by its antiparticle counterpart *and* reflecting the whole interaction in a mirror. According to CP symmetry, the mirror-image world will behave in exactly the same way that the real world does. But in a long series of experiments, beginning in 1963 and involving studies of the decay of particles known as K mesons (or kaons) James Cronin and Val Fitch of Princeton University discovered that roughly two decays in every thousand occurred in a way that violated CP symmetry. These decays only involve the weak interaction, but they show that the cherished principle of symmetry in particle-antiparticle interactions is not an absolute law of the Universe. This encouraged Sakharov to suggest, in 1967, that there must be processes involving the strong interaction and baryons that also violated the symmetry of particle-antiparticle interactions. If that were so, then he could sketch out a way in which baryonic matter could have been produced in the very early Universe.

All these ideas were greatly strengthened in 2004, when an experiment known as BABAR, which ran at the Stanford Linear Accelerator Center in the California, measured the decay processes of particles known as B mesons and their antiparticle counterparts. If there were no difference between the way fundamental interactions affected matter and antimatter, the two kinds of particle would decay in the same way, statistically speaking. But after sifting through the records of the decay of 200 million pairs of B and anti–B mesons, the researchers found that on 910 occasions a B meson decayed to a kaon and a pion, but on only 696 occasions did an anti–B meson decay in the same way. In the original kaon experiments, CP violation showed up only for two decays in a thousand, or 0.2 percent of the time; in the new experiments, it shows up at the 13 percent level (because there were 1,606 decays altogether, and a difference of 214 between the two decay modes; 214 is 13.3 percent of 1,606).* This is the strongest evidence yet that Sakharov's ideas about how matter emerged from the Big Bang are correct.

His suggestion looks so simple, with hindsight, that it is almost a tautology. But it required a completely different way of thinking about the Universe that nobody else had the imagination to come up with in the late 1960s. First, said Sakharov, there must be processes, operating at energies far greater than those achieved in accelerator experiments on Earth (which is why we have never seen them), which produce baryons (as opposed to antibaryons) out of energy. Second, at least some of those processes must violate CP symmetry. If not, there would be equivalent anti-processes that made antibaryons in just the right

---

*It's worth noting that the BABAR experiment alone, just part of the research going on at SLAC, involves some 600 scientists and engineers, from Canada, China, France, Germany, Italy, the Netherlands, Norway, Russia, the United Kingdom, and the United States. We won't continue to belabor the point, but this emphasizes the way in which science, as we mentioned earlier, is now a team game, not an individual pursuit.

numbers to annihilate the baryons made by the first process. Third, the Universe cannot be in a state of equilibrium (in effect, at the same temperature all the time) or the reverse processes would turn matter back into radiation as quickly as radiation turned into matter; this means that the Universe must be cooling, which in turn means it must be expanding. It is the expansion of the Universe which makes it possible for matter to freeze out from energy, provided there is an imbalance which produces more baryons than antibaryons.

Nobody took much notice of Sakharov's ideas at the time, because there was no detailed framework of theory and experiment on which to hang them. But as the models we described in Chapters 1 and 2 were developed in the 1970s, the idea surfaced again in the context of grand unified theories, and in particular the processes involving X bosons that imply the possibility of proton decay. Proton decay involves baryons disappearing from the Universe, converting matter particles into energy. Run this scenario backward, and you have an image of baryons appearing in the Universe, out of energy.

We know how much of the primordial energy of the Universe was turned into baryonic matter from two pieces of evidence. The first is simply a comparison of the amount of stuff we can see in stars and galaxies with the intensity of the background radiation. The radiation fills space uniformly, and this can be expressed as a radiation density, in terms of the number of photons in every cubic centimeter, or whatever volume you choose to work with. Baryonic matter fills the Universe less uniformly, but it is still possible to take the typical mass of a star, multiply it by the number of stars in a typical galaxy, count the number of galaxies in a chosen volume of space, and convert this into the density of baryons we would have if all the matter were spread out uniformly. As we shall see in the next chapter, there is slightly more to the story than this, since there is also a measurable amount of dark baryonic matter associated with galaxies; but the principle is straightforward.

The other approach depends on our understanding of the way protons and neutrons were "cooked" in the later stages of the Big Bang, and we shall go into that later in this chapter. Happily, both approaches give us the same answer for what is sometimes called the baryon to photon ratio—there is just one baryon in the Universe today for every billion (10^9) photons. (All such numbers are approximate, of course; nobody would worry too much if more accurate studies show that the ratio is a bit less or a bit more than this.) This is a measure of the size of the deviation of the decay processes involving X bosons from perfection—just one part in a billion. All of the GUTs predict that there will be such deviations from symmetry, but some predict larger values and some predict smaller values. One of the first triumphs of the marriage between cosmology and particle physics was the elimination of all the GUTs that do not predict roughly the right baryon to photon ratio; in particular, the exact size of this number strongly favors models incorporating supersymmetry.

This is the last piece of evidence we need in order to explain how the matter we see today was produced out of energy in the Big Bang. It starts with things we *think* we know about, the decay of X bosons, and ends with things we are sure we *know* about, the fusion of hydrogen nuclei to make helium nuclei in the last stages of the Big Bang fireball a couple of minutes later.

In the first instants of the Big Bang, pairs of X and anti-X particles were constantly being made out of pure energy in the usual way, and almost instantly they began interacting with one another and disappearing back into energy. But the mass of an X particle is 10^{15} GeV, and by 10^{-35} seconds after the birth of the Universe the temperature was already falling below the threshold at which pairs of X and anti-X could be made. At that time, there were still many such pairs around, but for every X there was an anti-X somewhere nearby. If all the surviving X and anti-X particles had met their counterparts and been annihilated, there would have been no baryons left over from the Big Bang to make stars, planets, and people. But the GUTs tell us that X bosons can decay in just the right way, thanks to CP violation and the expansion of the Universe, to leave a trace of quarks and leptons. In fact, because it has such a large mass, even a single X particle will decay into a shower of quarks and leptons. To keep things simple, however, we will just describe the basic process.

An X particle can follow either of two decay paths. Along one path, quark-antiquark pairs are produced, they annihilate one another, and nothing interesting happens. Along the other path, the pairs produced consist of an antiquark and a lepton, which go their separate ways. But this is not the end of the story. The anti-X particles are also decaying, following either of the equivalent two paths. Either they produce quark-antiquark pairs, of which no more need be said, or they produce pairs consisting of one quark and one antilepton — the opposite of the products from X decay. Once again, thanks to the expansion and cooling of the Universe, the end products of all these decays were left behind when the Universe became too cool to make new X particles. This is the relevance of Sakharov's crucial insight about the importance of the Universe's being in a state of nonequilibrium.

If this were all that was going on, after the X particles had all been used up, the particles produced by X decay would meet up with the antiparticles produced by anti-X decay, and vice versa, so once again all the matter would be converted back into energy. But CP violation tells us that matter and antimatter do not always behave in exactly the same way. In particular, the models based on observations of CP violation tell us that when all the X and anti-X particles have decayed, there will be a tiny bit more matter (one part in a billion) than there is antimatter. So when all the matter-antimatter pairs have annihilated each other, there will still be a trace of matter left in a Universe filled with radiation—just enough, if you choose the right model, to explain the observed baryon to photon ratio. One of the great hopes of particle physics is that the LHC and its associated experiments, including experiments with antimatter, may be able to test these ideas further. But we already have enough information to proceed to the next stage of the story of the development of the Universe, the processing of quarks into hydrogen and helium.

By the time the X particles had decayed, some 10^{-35} seconds after the birth of the Universe, the strong force, like gravity, had become a distinct entity. But at the high energies which still existed in the Universe there was no distinction between the electromagnetic and weak interactions. The behavior of particles was governed by three interactions (strong, electroweak, and gravity), and the particles we know as the carriers of the weak interaction, the W and Z particles, could roam freely through the Universe. Quarks (and, indeed, leptons) could still be produced in particle-antiparticle pairs out of energy, but from now on there would always be a tiny excess of matter over antimatter, left over from the decay of the X particles. *Individual* quarks would not survive from that era to the present day, but if one of these "original" quarks happened to meet up with a "new" antiquark and annihilate it that would leave the partner of that antiquark free, and so on down the generations as the Universe expanded.

Quarks themselves came into their own about 10^{-10} seconds after the beginning, when the temperature of the Universe fell below 100 GeV, the threshold required to make pairs of W and Z particles. From that time onward, the W and Z adopted their role of carrying the weak interactions between particles and had no independent existence except when they were produced (briefly) in high energy events involving collisions between particles, either naturally or in particle accelerators designed for the purpose. By now, the forces of nature had taken on their familiar roles as four distinct entities, with electromagnetism differentiated from the weak interaction.

The next phase of the development of the Universe involved a sea of hot quarks interacting with one another in a state known as the quark plasma. Some of the latest accelerator experiments are just beginning to probe the kind of conditions that existed in the Universe between about 10^{-10} and 10^{-4} seconds after the beginning, by smashing not just individual particles but beams containing nuclei of heavy elements such as gold and lead head-on into one another. As yet, little is known about the behavior of a quark plasma; but it is clear that between about 10^{-6} and 10^{-3} seconds after the beginning (that is, when the Universe was between 1 microsecond and 1 millisecond old), the temperature fell to the point where quarks no longer had enough energy to roam freely, instead becoming bound up in pairs and triplets, the way they are today. Starting at about 1 microsecond after the beginning, when the energy available had fallen below a few hundred MeV, quarks and antiquarks coalesced into baryons and antibaryons. In round terms, we can say that the quark plasma phase ended 10^{-4} seconds after the beginning; certainly by the time the Universe was a millisecond old all free quarks had disappeared. And still the tiny excess of matter over antimatter, inherited from the decay of X particles but now represented as a tiny excess of protons over antiprotons and neutrons over antineutrons, persisted as the Universe at last moved into an era in which those baryons became important constituents of matter. Most of the baryons annihilated with their antimatter counterparts to produce the sea of photons that still fills the Universe; the rest began the processes that led to our existence, in what we might call the baryonic phase of the Universe's existence.

It's worth pausing a moment here to think about the timescales involved. When we bandy about numbers like 10^{-10} and 10^{-35}, the natural reaction is just to see them both as tiny numbers. But 10^{-10} is 10^{25} times (that is, ten million billion billion times) bigger than 10^{-35}. In this sense, the era of inflation was as remote from the era of the quark plasma as we are from the era of the quark plasma, but on the other side. Which is why we only *think* we know what happened then. But at last we are ready to pick up the story of how the Universe developed from the time its density had fallen to that of nuclear matter today, and here we know exactly what was going on.

Before that time, up until about one-ten-thousandth of a second after the beginning, protons and neutrons were not the only baryons in the cosmic fireball. Heavier, unstable baryons could still be manufactured (in particle-antiparticle pairs) out of the energy available before annihilating each other once again. But as the temperature fell, no more of these heavier baryons could be manufactured, and the ones that were left either annihilated each other or decayed, ultimately into protons and neutrons. The threshold temperature for the pair-production of protons and neutrons themselves is about 10^{13} K, and the temperature of the Universe had fallen to about 10^{12} K by the time nuclear density was reached, about 10^{-4} seconds after the beginning.

There was, though, still plenty of energy available to make the much lighter electrons and positrons, so the image you should have is of a fireball with nuclear density but mainly consisting of photons and electron-positron pairs, laced with about one proton or neutron for every billion photons (and with an equivalent excess of electrons over positrons). At this stage, the number of neutrons was roughly equal to the number of protons, because of reactions involving neutrinos. At temperatures above 10^{10} K (ten billion degrees), a neutrino striking a neutron will convert it into a proton plus an electron, while an electron striking a proton will convert it into a neutron and a neutrino, and both reactions proceed with equal ease. But as the temperature dropped below ten billion degrees, which happened when the Universe was just over a second old, the fact that neutrons are a little bit (one-tenth of 1 percent) heavier than protons began to be important. With less and less energy available, it was increasingly difficult to make up the mass difference when an electron hit a proton, so the reactions that made neutrons out of protons became less effective than the reactions that make protons out of neutrons, where no extra input of energy is required. By one-tenth of a second after the beginning, the proportion of neutrons to protons had dropped to 2:3; about a second after the beginning, the number of neutrons had dropped even further, so that only a quarter of the mass in baryons was in the form of neutrons — in other words, there was one neutron for every three protons. The neutrons might have disappeared entirely, but at about the same temperature, 1 MeV, the influence of the neutrinos, which operate only through the weak interaction, became less effective.

Remember that a second after the beginning, when the temperature was about a billion degrees, the conditions in the entire Universe resembled those at the heart of an exploding supernova today. Under the conditions of extreme pressure, density, and temperature that exist in the heart of a supernova, neutrinos still interact strongly with baryonic matter. But neutrinos produced by particle interactions at the heart of an ordinary star like the Sun stream out through the whole body of the star more easily than light passes through a sheet of clear glass. From the time when the Universe was about a second old, neutrinos essentially ceased interacting with protons and neutrons, except in occasional, rare collisions. The Universe became transparent to neutrinos when its density fell below about 400,000 times the density of water, and they are said to have "decoupled" from everyday matter. But the neutrinos are still here — an estimated billion or so of them in every cubic meter of space, or several hundred in every cubic centimeter — and may still be important in other ways, as we shall see.

Even after the Universe was a second old, occasional interactions involving electrons with higher than average energy were still capable of making neutrons out of protons, although the number of such interactions was falling rapidly. By 13.8 seconds after the beginning, when the temperature had fallen to three billion degrees, 17 percent of the baryons were still in the form of neutrons. This moment is important in the story of the development of the Universe, because at three billion K there was no longer enough energy even to make electron-positron pairs, and the ones that were left gradually annihilated one another after this time, leaving behind the trace of electrons ultimately derived from the breaking of CP symmetry, which exactly balanced the number of protons, with one electron left over for every proton in the Universe (in this sense, a neutron counts as a combination of a proton and an electron, produced by those interactions involving neutrinos). No longer bathed in a sea of energetic electrons and positrons, the remaining protons and neutrons were essentially left to their own devices.

Left to their own devices, protons, as we have seen, are very stable and long-lived. But a lone neutron is unstable and will decay into a proton, an electron, and an antineutrino, with a half-life of 10.3 minutes. This means that however many neutrons you start out with, after 10.3 minutes half of them will have decayed; in every 100 seconds, about 10 percent of the free neutrons will decay. But long before the Universe was 10.3 minutes old, the surviving neutrons had been locked up safely in atomic nuclei, where they are stable and do not decay.

Apart from the proton itself, which can be regarded as the nucleus of a hydrogen atom, the simplest atomic nucleus is that of deuterium, and consists of a single proton and a single neutron bound together by the strong force. Such nuclei began to form briefly around this time, when the Universe was less than half a minute old, but were soon knocked apart in collisions. The binding energy of a deuteron (as a deuterium nucleus is also called) is only 2.2 MeV, which means that any impact with another particle carrying that much energy (a proton, a neutron, or even a suitably energetic photon) will break it apart. By the time the Universe was a hundred seconds old, the proportion of neutrons had fallen to about 14 percent—there was just one neutron left for every seven protons. But at this time the temperature of the Universe fell below a billion degrees (a little less than a hundred times the temperature at the heart of the Sun today), corresponding to particle energies of only about 0.1 MeV. There was no longer enough energy in particle collisions to break deuterons apart, and any neutrons that became bound up with protons in this way were safe from decay.

This process, known as nucleosynthesis, didn't stop there. As the energy of impacts dropped, deuterons themselves were involved in further interactions with neutrons, protons, and other deuterons, making slightly heavier nuclei. Adding an extra neutron makes a nucleus of tritium (two neutrons and one proton), while even more stable nuclei are composed of two protons plus one neutron (helium-3) or two protons plus two neutrons (helium-4). The most stable of all these nuclei is helium-4, which has a binding energy of 28 MeV, corresponding to 7 MeV for each baryon in the nucleus. Because of its stability, almost all of the available neutrons were quickly locked up in nuclei of helium-4, although there were also tiny traces left over of deuterium, tritium, helium-3, and one slightly heavier nucleus, lithium-7, which contains three protons plus four neutrons and is essentially made by sticking together one helium-4 nucleus and one tritium nucleus. But no heavier elements were manufactured in the Big Bang, because the temperature of the Universe soon fell below the point where nuclei could be stuck together in more complicated ways. Because nuclei and protons all have positive electric charge, and like charges repel, it takes a certain amount of energy to squeeze them closely enough together for the short-range strong force to take over and pull them together; by the time the Universe was a couple of hundred seconds old there just wasn't enough energy left to overcome this electrical barrier.

At the time all this happened, there was one neutron in the Universe for every seven protons — one in eight baryons were neutrons. As each helium-4 nucleus contains the same number of protons as neutrons, this means that one proton was locked up for every neutron; overall, two baryons in eight, or one-quarter of the total number, were locked away in helium-4, and three-quarters of the baryons were left as free protons to form hydrogen nuclei, except for the tiny traces of the elements we have mentioned, which together add up to just a fraction of 1 percent. As the masses of protons and neutrons are roughly the same, this means that one-quarter of the mass of baryons that emerged from the Big Bang was in the form of helium, and three-quarters of the mass was in the form of hydrogen. Most of this activity was completed by the time the Universe was four minutes old; nucleosynthesis stopped entirely by the time the Universe reached the ripe old age of thirteen minutes. At that time, though, there were still no atoms, just free nuclei and free electrons moving in a sea of radiation that was still highly energetic by terrestrial standards. Nothing much happened for the next few hundred thousand years, except that the Universe, still dominated by radiation, continued to expand and cool.

The Universe really was still *dominated* by radiation at this time. Around the time nucleosynthesis began, 0.1 seconds after the beginning, the density of the Universe was between five and ten million times the density of water. But the fraction of that density contributed by the baryons on their own was only equivalent to about one and a half times the density of water. Almost all of the rest was the density contributed by the energy of radiation—the density of photons, if you like—in accordance with $m = E/c^2$. Today, although matter is spread pretty thin, the dynamics of the Universe are dominated by the influence of matter—with one proviso that we discuss in later chapters but which does not affect the discussion here—and the radiation is reduced to the weak hiss of the microwave background, with a temperature of just 2.73 K. After nucleosynthesis, the next important landmark in the development of the Universe occurred when radiation became less important than matter in terms of density. It happened a few hundred thousand years after the beginning, thanks to a key difference in the way matter and radiation behave when they are compressed or when they expand.

Density is the measure of the amount of stuff in a certain volume. In a (flat) three-dimensional Universe, the volume of a region of space is proportional to the cube of its linear size—a sphere that is twice as big in radius as another sphere has eight (2^3) times the volume of the smaller sphere. So when the present-day observable Universe was half its present linear size, with all the galaxies separated from one another by only half their present distances, it had one-eighth of its present volume and the density of matter in it was eight times the present density. But radiation density obeys a slightly different rule. If you imagine a box of radiation and double the length of each side of the box, the volume increases by a factor of eight, and the density of the radiation falls by a factor of eight in the same way as for matter. But at the same time, the wavelength of the radiation gets stretched by a factor of two — the famous redshift, so called because in the rainbow spectrum of light red has a longer wavelength than other colors. This corresponds to a weakening of the energy of the radiation, which means a decrease in its mass equivalent. So overall, the change in energy density does not go as the cube of the change in linear dimensions but as the fourth power. When the present-day Universe was half its present linear size, the density of radiation in it was not eight times greater than it is today but sixteen (2^4) times greater. When the Universe was, in a linear sense, one-tenth its present size, the baryonic density was a thousand times greater than today, but the radiation density was ten thousand times greater than today, and so on. It is easy to see how this process enhances the importance of radiation as we look back in time, until we find that a few hundred thousand years after the beginning, matter and radiation made equal contributions to the density of the Universe, while at earlier times radiation was by far the more important constituent.

At the same time that the density of radiation dropped below the density of baryons, when the Universe was a few hundred thousand years old, matter and radiation "decoupled" from each other and went their separate ways. Before that time, the temperature was too high to allow electrically neutral atoms to form. But photons — which are, after all, the carriers of the electromagnetic interaction — interact powerfully with charged particles. The positively charged nuclei and negatively charged electrons moved in a hot sea of photons, forming a plasma, in which the photons interacted with charged particles (in effect, bouncing off them) at every turn, following a zigzag path through space like a high-speed ball in a crazy cosmic pinball machine. Provided the temperature of the Universe were more than a few thousand degrees, any electron that was captured by a nucleus would almost instantly be knocked free by the impact of an energetic photon. But as the temperature fell below this threshold, the impacts from photons became too weak to break the bonds of the electromagnetic forces holding atoms together, and all the electrons and nuclei gradually got locked up in neutral atoms. With no more free electrically charged particles around to obstruct them, the photons were able to stream through space largely uninterrupted.

It is no coincidence that all this occurred at about the temperature of the surface of the Sun today, because exactly the same process is going on there now. Below the surface of the Sun, where the temperatures exceed 6,000 K, electrons are stripped from neutral atoms by energetic impacts, and the matter is in the form of a plasma similar to the last stages of the fireball in which the Universe was born. To give you some idea of how difficult life is for a photon trapped inside such a plasma, a photon starting out from the heart of the Sun travels, on average, just 1 centimeter before it collides with a charged particle and bounces off in a random direction. So it moves in zigzag steps each about a centimeter long, and typically takes ten million years to reach the surface, even though it is traveling at the speed of light. If it could go in a straight line from the center of the Sun to the surface, its journey would take just 2.5 seconds. But it has literally traveled a total of ten million light years — backward, forward, and sideways in centimeter steps — on its way out. If the zigzag path were straightened out, it would stretch five times farther than the distance from here to the Andromeda Galaxy, the nearest large neighbor of our Milky Way. It is only at the surface of the Sun, where electrons combine with nuclei to make neutral atoms, that the photons can stream freely out into space.

Because we live on a planet where electrically neutral atoms are normal, and they have to be broken apart to make a plasma which can then recombine to make atoms, physicists call the process by which the nuclei and electrons in a plasma get together to make neutral atoms recombination. They even apply this term to the events that took place when the Universe was a few hundred thousand years old, although, strictly speaking, this was not "re" combination but "combination" — the first time in the history of the Universe that electrons and nuclei got together in this way. Whatever it's called, at recombination the entire Universe resembled the surface of the Sun today, and the photons we detect today as the cosmic microwave background radiation have streamed through space ever since then without interacting with any matter at all until they fell into the dishes of our radio telescopes.

There's a neat analogy to give you some feel for just how far back toward the Big Bang these radio telescopes are looking. It comes originally from the American physicist John Wheeler but was updated by Alan Guth in his book *The Inflationary Universe.* If our view back in time across the Universe is likened to the view down to the street from the top of New York's Empire State Building, with street level representing the beginning, 14 billion years ago, then the farthest galaxies seen to date correspond to the tenth floor above street level, and the most distant quasars yet observed are at the equivalent of the seventh floor. But the era of recombination glimpsed in the form of the background radiation corresponds to a view of something just a centimeter above street level. That is why observations of the background radiation are so important for our understanding of the early development of the Universe.

Even without going into the importance of the tiny fluctuations in the temperature of the fireball from place to place, just measuring the overall temperature of the background radiation today and knowing the "density" of photons in the Universe provides key insights into the nature of the Universe. In the story outlined in this chapter, we have identified temperatures (energies) at different times during the development of the Big Bang. But how do we know those temperatures with such accuracy? Simply because we can measure the temperature of the background radiation today, and then use the equations that describe what happens to radiation when it is squeezed plus the equations of the general theory of relativity that describe how the Universe expanded, to work backward in time to any epoch of interest. This tells us the temperature at, for example, the time of primordial nucleosynthesis (or, if you prefer, it tells us at what time in the development of the early Universe the temperature was right for nucleosynthesis to occur).

The rate at which that primordial nucleosynthesis proceeded, however, didn't depend only on temperature. It also depended on the density of baryons (specifically, the nuclear particles, protons and neutrons, that are collectively known as nucleons) at that time. The more nucleons there were, the more nuclear interactions could take place; the fewer the nucleons, the less likely the reactions that built up deuterium, helium, and lithium would be. Because we know the density of photons in the Universe rather well, it is convenient to measure the density of nucleons in relative terms, compared with the density of photons. Different nuclear reactions depend on this ratio with different degrees of sensitivity. The most sensitive are the reactions that produce deuterium. The calculations show that if there were one nucleon for every 100 million photons at the time of primordial nucleosynthesis, then there would be only 0.00008 parts per million of deuterium in baryonic material today — just eight nuclei in every 100 billion would be deuterium. If the photon to nucleon ratio is a billion to one, there would be sixteen deuterium nuclei per million. And if the ratio were 10,000 million to one, there would be six hundred deuterium nuclei per million. In fact, spectroscopic observations of the oldest stars show a deuterium abundance of between sixteen and twenty per million nuclei, corresponding to a photon to baryon ratio of just over a billion to one.

Spectroscopy is such a key tool of astronomy that it is worth a brief digression. Since every kind of atom (every element) produces a characteristic imprint of lines at particular wavelengths in the spectrum of light, as unique as a fingerprint or a supermarket barcode, astronomers can tell what something is made of as long as they can see light from it, even if it is far away across the Universe. Because cold atoms absorb particular wavelengths of light with exactly the same pattern they would radiate if they were hot, we can also find out what cold clouds of gas and dust in space are made of by analyzing light which has passed through them from distant stars. When objects move toward us through space, the whole pattern of spectral lines gets squashed up toward the blue end of the spectrum, and when they move away from us the pattern gets stretched out toward the red end of the spectrum (a redshift). This Doppler shift tells us how fast stars and galaxies are moving through space. The famous cosmological redshift is produced by a different process, the stretching of space itself between galaxies; it tells us how fast the Universe is expanding and, by implication, when it was born. Without spectroscopy, we would know very little about the Universe we inhabit, and this book (among many others) would never have been written.

With the aid of spectroscopy, though, we can measure the proportions of helium and lithium, as well as of deuterium, in old stars, and use these measurements to refine our calculation of the density of nucleons at the time of primordial nucleosynthesis. It is because we know these ratios — for example, the fact that some 25 percent of the baryonic mass is in the form of helium — that we know what the conditions were like during the era of primordial nucleosynthesis. All the numbers agree, provided that the density of nucleons in the early Universe was somewhere in a rather small range of values. In terms of grams per cubic centimeter, the numbers are too small for easy comprehension — they correspond to a density of baryonic matter in the Universe today of a few times 10^{-31} grams per cubic centimeter. It makes more sense to think in terms of the critical density for which the Universe would be exactly flat.

As we have seen, there is every reason to think that the Universe is flat in this sense, and inflation theory predicts that it must be indistinguishably close to being flat. Astronomers define this critical density for flatness as 1. When I was starting out in astronomy in the late 1960s and 1970s, the evidence from the background radiation and the observed abundances of the lightest elements in the oldest stars told us that the density of baryons in the Universe today lay between 0.01 and 0.1. That is, baryons — the stuff we are made of, and the stuff all the bright stars and galaxies in the Universe are made of — contribute between 1 percent and 10 percent of the mass needed to make the Universe flat. At the time, this discovery seemed (and was) an astonishing achievement of science and the human mind. But by 2005 improved observations had shown that there is between 4 percent and 5 percent, and probably nearer 4 percent, of the critical density in the form of baryons — the measurements are at least ten times better than they were thirty years ago. (And no more than a fifth of this baryonic matter, less than 1 percent of the mass needed to make the Universe flat, is in the form of bright stars and galaxies.) There is no escape from this conclusion, so if the Universe really is flat then it must contain some form of matter which is not made of baryons (that is, nonbaryonic matter) and which cannot be seen, because it does not shine (in other words, dark matter, or possibly dark energy). At least 95 percent of the mass of the Universe must be nonbaryonic.

This, it turns out, is a good thing. Although the nature of this dark stuff remains mysterious (as we shall see in Chapter 6), without it there would be no galaxies and possibly no stars. The gravitational influence of the dark stuff has been crucially important in the development of the kind of structure we see in the Universe today (including ourselves), starting out from the small fluctuations in the cosmic fireball which are revealed by the ripples in the background radiation, and which themselves originated in quantum fluctuations taking place during the epoch of inflation.

5

Why Is the Universe the Way It Is?

At the time of recombination, a few hundred thousand years into the life of the Universe, the irregularities in the distribution of baryonic matter, revealed by the ripples in the cosmic microwave background radiation, amounted to only one part in a hundred thousand. That is equivalent to ripples just 1 centimeter high on the surface of a lake 1 kilometer deep. If there had been nothing but baryons in the Universe, providing only 5 percent or less of the density needed for flatness, the expansion of the Universe would have stretched those ripples and pulled them apart before gravity could tug the baryons together to make interesting things like stars and galaxies. The gravitational pull of a ripple that small would have been simply too feeble to resist the cosmic expansion. But there is other evidence, also from the analysis of the background radiation, that within a billion years — perhaps as soon as half a billion years — from the time that background radiation was released, hot objects, something like stars or superstars, had formed and were exerting an influence on their surroundings.

The relevant influence was that these first superstars made the gas near them hot. This heating stripped away electrons from atoms of hydrogen and helium, reionizing the material that had combined into the form of neutral atoms at recombination, when the Universe was roughly a thousandth of its age at the time of reionization. This meant that once again there were free electrons in the Universe that could interact with the electromagnetic radiation left over from the fireball of the Big Bang. But because the density of the Universe had dropped dramatically by this time, the background radiation was not completely obscured. Instead, the ionized material left an imprint on the radiation as it passed through. Indeed, the imprint we see (which affects the polarization of the radiation, in the same sort of way that polaroid sunglasses affect light passing through them) is a result of interactions between the radiation and free electrons all along the line of sight from the time of reionization to the present day, a low-density "column" of material roughly 13 billion light years long. The observations tell us roughly how many electrons there are in such a column in any direction on the sky. Combined with our understanding of how the density of baryons in the Universe has changed as the Universe has expanded, this indicates how long the column must be, and therefore where (or when) the end of it is at the time of reionization.

There is some uncertainty in the estimates, partly because reionization probably happened over a period of hundreds of thousands of years, rather than switching on everywhere in the Universe at once, and partly because observations based on studies of the microwave background give slightly different estimates from those based on studies of the most distant bright objects known, which are called quasars. But these are details which will be resolved by the next generation of detectors. What is unambiguously clear is that reionization had occurred by a billion years after the Big Bang, at distances from us that correspond to redshifts greater than 7. (Because light takes a finite time to cross space, and the redshift of a cosmological object tells us its distance in the expanding Universe, redshifts can be translated into "look-back time." The look-back time for a redshift of 6, for example, corresponds to a look-back time of 12.5 billion years. Recombination occurred at a redshift of 1,000.)

Confirmation that matter had already clumped together to make stars and small galaxies (called dwarf galaxies) by this time has come from the Hubble Space Telescope, which took a long-exposure photograph of a tiny patch of sky to record images of the faintest and most distant objects in the field of view. This is called the Hubble Ultra Deep Field, or HUDF. Analysis of the HUDF in 2004 showed roughly a hundred faint red smudges, each one corresponding to a dwarf galaxy, seen by light which left it when the Universe was little more than a billion years old. But even these cannot be the first hot objects, which must have formed at redshifts of around 15 to 20, corresponding to a look-back time of 13 billion to 13.5 billion years, only a couple of hundred million years after the Big Bang. If we could image those objects, this would be equivalent to a seventy-year-old person "looking back" at photographs of him- or herself as an eleven-month-old baby.*

There is still an element of guesswork, aided by simulations made

---

*To the surprise and delight of many astronomers, late in 2005 a team of observers studying infrared light detected by the Spitzer space telescope reported that they had seen a faint glow of background radiation at infrared wavelengths (quite distinct from the cosmic microwave background radiation), which could be the smeared out and highly redshifted light from these primordial "Population III" stars, seen across 13 billion light years of space. This is independent evidence in support of the calculated time of reionization, but still not the same as being able to image individual stars directly from that long ago.

on computers, about how the first hot objects formed, but we *think* we know what happened. The following scenario is probably right at least in outline and will be tested and improved by the next generation of space telescopes.

The first requirement is to establish what kind of dark matter allows galaxies to form. Although we shall be saying a lot more about the dark stuff that holds the Universe together in the next chapter, there is one key distinction that needs to be made here. When astronomers first appreciated the need for dark matter to explain the dynamics of the Universe, there were two kinds of candidate for the role and one known contender. The contender was the neutrino. Neutrinos had always been assumed to have zero mass, and therefore (like photons) to travel at the speed of light. But at least physicists knew that neutrinos existed, and there are so many neutrinos in the Universe (roughly the same number of neutrinos emerged from the Big Bang as the number of photons in the background radiation) that even if each of them had only a very small mass, that mass would add up to a significant fraction of the density required for flatness. As late as the beginning of the 1990s, no solely Earth-based experiment had been able to establish whether neutrinos did have mass, but a surprising amount of information about neutrinos had come in from astronomical and cosmological studies.

Leaving aside the mass question for the moment, it was actually cosmology that first established that there are precisely three varieties (three "flavors") of neutrino, the respective counterparts to the electron, the tau particle, and the muon. Neutrinos are so elusive that *proving* other varieties do not exist is a difficult task, which depends on indirect arguments and a lot of guesswork. From terrestrial experiments alone, in the early 1980s all that physicists could say was that there must be fewer than 737 flavors of neutrino, and over the next few years they struggled to get the limit down to 44 flavors, then 30 flavors, and by the second half of the 1980s, thanks to a major effort at CERN, down to 6 flavors. But all they were doing was confirming what the cosmologists already knew.

The reason is that the number of flavors of neutrino influences the amount of helium produced in the Big Bang at the time of primordial nucleosynthesis. The exact amount of helium produced depends on how fast the Universe is expanding at the time of nucleosynthesis, in such a way that the faster the Universe is expanding the more helium there is (because there is less opportunity for free neutrons to decay before getting locked up in helium nuclei). One of the things that affects the expansion rate is the number of varieties of light particles present (and their antiparticles, which will be taken as read). You can think of this as a kind of pressure aiding the expansion of the Universe, with more kinds of light particles producing more pressure and making it expand faster. The cosmological calculation tells us that for the observed abundance of just under 25 percent helium seen in the oldest stars, there can only have been five varieties of light particle in the Universe at the time of primordial nucleosynthesis. Two of these were the photon and the electron, which leaves room for just three varieties of neutrino. For each additional flavor of neutrino, the amount of helium would rise by one percentage point; having four flavors would push the amount of helium above 25 percent, and this is ruled out by the astronomical observations. Since the 1980s, the accelerator experiments on Earth have become powerful enough to set the same limit on the number of neutrino flavors. From that perspective, the measurement of the number of neutrino flavors here on Earth tells you how much helium was made in the Big Bang — striking confirmation that both particle physics and cosmology are dealing with fundamental truths about the nature of the Universe.

Astronomical observations were also the first to tell physicists that neutrinos have mass, from studies of something much closer to us than the Big Bang—the Sun. And this provides another link between laboratory scale physics, astrophysics, and cosmology, reinforcing the point that the underlying science is something we really *know* about how the world works.

The story, which we referred to in the context of grand unified theories in Chapter 2, goes back just over forty years, to the early 1960s. That was when a team from the Brookhaven National Laboratory, headed by Ray Davis, installed an experiment 1.5 kilometers below ground in a mine at Lead, South Dakota, to detect neutrinos from the Sun. The experiment had to be buried so deep to avoid interference from particles from space known as cosmic rays; but it also had to be very sensitive, since neutrinos are extremely reluctant to interact with matter at everyday densities. They stream out from the heart of the Sun without being deflected on the way (unlike the photons following their tortuous zigzag paths to the surface) and, indeed, pass easily through the 1.5 kilometers of solid rock above the detector. The detector itself was one-fifth the size of an Olympic swimming pool, a tank filled with 400,000 liters of perchlorethylene, a fluid commonly used in dry-cleaning processes. Rare interactions between neutrinos from the Sun and chlorine atoms in the perchlorethylene were expected to produce atoms of a radioactive isotope of argon which could be counted by suitable instruments.

The processes going on inside the Sun that produce neutrinos involve the conversion of hydrogen into helium, reminiscent of the process of primordial nucleosynthesis in the Big Bang. This fusion process releases energy—which is what keeps the Sun shining—and neutrinos. Because we can measure how much energy is flowing out from the Sun, and we know from laboratory studies how much energy is produced when one helium nucleus is made, we can calculate how many nuclear reactions take place each second and how many neutrinos are produced as a result. The calculation tells us that seven billion neutrinos of the kind that could interact with chlorine atoms in the right way cross every square centimeter of the Earth (including the detector) every second. But in a sign of just how feebly neutrinos interact with matter at everyday densities, the calculations predicted that just twenty-five neutrinos would be detected at the Homestake Mine every *month*. In fact, over several decades the experiment recorded a flux of just one-third of the expected number of neutrinos, eight or nine a month instead of twenty-five.

Since the 1960s, many other detectors of different kinds have recorded similar results, and experiments involving the flux of neutrinos produced by nuclear reactors or by cosmic rays interacting with atoms in the atmosphere (no longer just electron neutrinos, but other kinds as well) have also shown related discrepancies between the number of neutrinos produced in the reactors or by the cosmic rays and the numbers recorded in the detectors. The only viable explanation for the discrepancy is that the neutrinos are being produced in the expected numbers, but something happens to them on their way to the detectors.

The neutrinos that are produced in the interactions at the heart of the Sun are all electron neutrinos, and the detectors used in the archetypal solar neutrino studies can only detect electron neutrinos. But it is now clear that as they travel through space the neutrinos change into the other varieties (tau and muon neutrinos) and back again. The process is called neutrino oscillation, and it means that if you start with a beam of pure electron neutrinos (or, indeed, one of the other varieties) you soon end up with a beam consisting of one-third electron neutrinos, one-third tau neutrinos, and one-third muon neutrinos.

The process is related to the quantum phenomenon of wave-particle duality, and it wasn't just pulled out of the hat to explain the solar neutrino puzzle — such oscillations were well known from studies of particles called kaons before they were invoked to account for the observations of neutrinos. But there is one absolutely crucial feature of such oscillations. They can only happen for particles that have mass. In other words, measurements made at a tank of cleaning fluid down a mine in South Dakota tell us that the most ubiquitous particles in the Universe, neutrinos, must have mass. The mass of each neutrino may be very small, but it cannot be zero.

You might think — many astronomers did, for a time — that this solves the puzzle of the "missing" mass in the Universe. But it quickly turned out that putting all the missing mass into neutrinos would not work. Neutrinos might be dark, but they are the wrong kind of dark matter to explain the origin of the observed pattern of the bright matter in the Universe. The two kinds of dark matter we referred to earlier are known as "hot" and "cold" dark matter. Neutrinos are hot, in the sense that they move around at a sizable fraction of the speed of light. But what we need to explain the pattern of galaxies on the sky is a profusion of particles of slow-moving cold dark matter, or CDM.

Remember that the puzzle we have to explain is how tiny ripples amounting to overdensities of just one part in a hundred thousand in the distribution of baryons in the early Universe grew to become galaxies and clusters of galaxies today, even though the Universe was expanding and trying to stretch them even thinner. The pattern made by the bright stuff of galaxies on the largest scales resembles that of the inside of a natural sponge, with holes (regions devoid of bright galaxies) surrounded by bubbles of bright stuff in the form of sheets and filaments made up of clusters and superclusters of galaxies. It is possible to make this kind of structure in a universe dominated gravitationally by hot dark matter, but the problem is that it takes a very long time. High velocity particles with sufficient mass emerging from the Big Bang would scatter clumps of baryonic matter as effectively as a bowling ball scattering the pins during a strike, and would sweep the baryons up into sheets and filaments around the edges of empty regions of space. The process would end only when the neutrinos had slowed down ("cooled") to about a tenth of the speed of light, and only then could the great sheets of hydrogen and helium gas begin to break up and collapse under the pull of gravity, eventually building stars and galaxies in a "top down" process.

But *eventually* is the key word. The whole process would take at least four billion years, but we know the Universe is only 14 billion years old, and there are stars older than ten billion years even in our own Galaxy, while deep surveys such as the Hubble Ultra Deep Field show that small galaxies (but not large ones) had already formed a billion years after the Big Bang. It turns out that astronomers would actually be embarrassed if neutrinos had more than a *very* tiny mass — but, happily, the experiments suggest that the masses of neutrinos are indeed too small to pose a problem for models of galaxy formation.

The reason why neutrinos have to have mass in order to oscillate is that the rate at which the oscillation occurs depends on the *difference* in mass between the different varieties of neutrino. If the masses were zero, there would be no difference. Because the rate of oscillation depends on the mass difference, the distance traveled by the neutrinos before they are fully mixed into an equal amount of all three varieties also depends on the mass difference. The solar neutrino studies alone don't tell us much about this, because the distance from the Earth to the Sun is so great that even light takes 8.3 minutes to travel from the Sun to the Earth, and neutrinos take slightly longer. This is an enormous time by the standards of most particle interactions, ample for complete mixing to occur. But the studies of neutrinos produced in the atmosphere by cosmic rays, which take only a fraction of a second to reach the detectors, set much more stringent limits. These studies cannot tell us directly what the masses of the individual verities of neutrinos are, but they can give an indication of the total mass of all three kinds of neutrino, assuming that they behave in a similar way to more easily studied particles such as kaons. Translating this into an overall contribution of neutrinos to the density of the Universe, tells us that neutrinos contribute at least 0.1 percent of the mass needed to make the Universe flat.

On the other hand, the kind of structure we see in the Universe today and the length of time for which that structure has been around tell us that the contribution of *all* kinds of hot dark matter to the density of the Universe is no more than 13 percent of the mass in baryons—in other words, in round terms, no more than 0.5 percent of the total density required for flatness. This is a satisfactory agreement between observations of the smallest, lightest things in the Universe and the largest structures in the Universe. It would have been decidedly embarrassing if the particle physics had said that neutrinos must contribute at least 0.5 percent of the flatness density but galaxy studies had said that the contribution could be no more than 0.1 percent; but the reverse is the case. It cannot be overemphasized that this kind of agreement is powerful evidence that physicists know what they are talking about, even if the numbers don't match precisely.

Translating this into energy and mass units, if you had one neutrino of each of the three varieties their total mass would add up to less than 2 eV, equivalent to 0.0004 percent of the mass of a single electron. So we *think* that adding the masses of all three kinds of neutrino together contributes between 0.1 percent and 0.5 percent of the gravitational stuff needed to make the Universe flat, and we still have to account for 95 percent of the gravitational stuff in the Universe. The first step is to look at how *cold* dark matter played a part in the growth of the structure we see in the Universe today — but without, for now, worrying about exactly what CDM particles might be, since on this occasion the cosmology came first, and told the particle physicists what to look for.

Astronomers test their ideas about the growth of structure in the Universe by comparing observations of the patterns traced out by galaxies and clusters of galaxies on the sky with the predictions of simulations of the growth of irregularities as a result of gravitational attraction in an expanding universe. Stated that baldly, it sounds simple. But the observations require measuring the redshifts of hundreds of thousands of galaxies, far too faint to be seen by the naked eye, on different patches of the sky. Such detailed studies became practicable only with digital technology — CCDs (charge-coupled devices) to "photograph" the galaxies, computers to analyze the data — at the end of the twentieth century and the beginning of the twenty-first. The redshift measurements are converted into distances to build up a three-dimensional map of a wedge or cone of the Universe extending outward from our point of view. Even today, this has not been done for the whole sky; but different observations of widely separated patches of the sky give the same kind of overall foamy picture, so we are confident that these slices represent the typical view of the Universe.

If anything, the computer simulations are harder. If you had a big enough computer (that is, one with a big enough memory), you could represent every galaxy in a model of the early Universe as a particle characterized by a set of numbers, plug in Einstein's equations for the universal expansion and the law of gravity, and run the model with different starting conditions and different amounts of cold dark matter to see which ones end up looking like the real Universe. With several hundred billion known galaxies, this is clearly impossible. Instead, each "particle" in such a simulation corresponds to about a billion times the mass of the Sun. In the largest of these simulations, ten billion virtual particles are used to simulate the behavior of the entire visible Universe as it expands.* Starting out with such particles distributed statistically in the same way as the distribution of matter at the time of recombination, the computer models can then step forward in virtual time to see how the particles cluster together. When things start to get interesting, the simulation can then focus on one of the clusters that is forming, ignore the rest of the model universe, and be recalibrated to use the same number of virtual particles to probe the development of structure within that cluster on smaller scales. In principle, the process could continue right down to the formation of individual galaxies, but this is pushing present-day computer technology to the limit.

Such studies are, like most modern research, far beyond the scope of individuals. The largest of these simulations is the work of an international group of scientists called the Virgo Consortium, named after the nearest large cluster of galaxies in the Universe, which lies in the direction of (but far beyond) the constellation Virgo. The calculation proceeds by choosing one of the virtual mass points and calculating the gravitational influence on it of each of the other 9,999,999,999 points, then choosing another point and doing the same thing, over and over again until every point is accounted for. Each point is moved a little bit in the simulation in accordance with all the gravitational

---

*A quick calculation will tell you that even ten billion such particles add up to only 0.003 percent of the mass of the visible Universe. This doesn't matter too much, because the Universe is so close to being smooth and homogeneous on the largest scales. In a similar way, a "map" of a regularly planted field of maize need only show every hundredth maize stalk to give you a good idea of what the field looks like, since the rows of maize are uniform.

forces; the "Universe" is expanded by a tiny bit; and the entire process is repeated yet again.

But in order to make progress in a reasonable time (that is, before the researchers die of old age), some shortcuts have to be used. For example, for points that are sufficiently far apart, the simulation combines thousands of individual particles and uses their total gravitational influence in calculating the effect on a particle on the other side of the model universe, instead of calculating all their individual contributions. The cluster of Unix computers used in this simulation incorporates 812 processors with 2 terabytes of memory performing 4.2 trillion calculations per second (4.2 teraflops). Even at that speed, the model runs for weeks at a time to produce results. By the middle of 2004, the simulation had produced 20 terabytes of data which represent 64 snapshots of the virtual universe at different stages in its development — different redshifts, looking back in time from today. Comparison of these snapshots with the patterns made by bright galaxies in the redshift maps of the real Universe clearly show that a substantial amount of dark matter must be present to explain the kind of structure we see in the real Universe.

Of course, these comparisons are not made by eye, even though a casual look at the two kinds of map does provide a striking impression of similarity. Instead, statistical comparisons of the kinds of sheets, filaments, and voids seen in the computer simulations (one of which provided the image used on the jacket of this book) and in the real Universe give an objective measure of just how well the simulations match up to reality. The answer is, very well indeed — *provided* that there is lots of cold dark matter and the Universe is flat.

Although many different simulations have been carried out, with different amounts of dark matter in them, different values for the density of the universe and the deviation from flatness, and so on, there is no point in going through all of them, since only one really matches the Universe we live in. But it wasn't a lucky guess, and we wouldn't want you to think that astronomers just pulled it out of the hat and found they were right; getting there involved a lot of false starts and backtracking out of blind alleys. The model we have now is the best we've got, and the best understanding of the Universe there has ever been, but it evolved out of decades of work, reminiscent of the way a modern jet aircraft has evolved from the Wright brothers' first flyer.

The model is based on the idea that baryons are embedded in a sea of cold dark matter. We shall say more about the nature of CDM in the next chapter, but what matters here is that it is required by the cosmology, and it seems to be in the form of particles which do not interact with baryonic matter in any way except through gravity. We cannot be sure how many of these particles there are, or what their individual masses might be (or even whether they come in more than one variety), but a reasonable guess is that they have the same sort of masses as protons and neutrons. The computer simulations show that they are spread through the entire Universe, including the voids between the bubbles of bright clusters of galaxies. There must also be dark baryons in these voids, since in order to make the simulations match up with the real Universe we have to assume that baryons and CDM particles are intermingled across the Universe. We only see a foamy pattern of bright galaxies because bright galaxies have only formed in regions where there is a slightly greater density of dark matter, which has pulled the nearby baryonic gas into gravitational potholes where gas clouds have become massive enough to collapse and form stars and galaxies. This means that the distribution of bright stuff gives us a slightly biased view of the Universe, where matter is actually distributed slightly more uniformly than the bright stuff. But this biasing is quite small—if, as seems to be the case, the average density of matter in the Universe is nearly enough to cause clouds of gas to collapse, it only needs a relatively small ripple of excess density to start the process.

The intimate relationship between baryonic matter and CDM is also shown by studies of individual galaxies like the Milky Way. Indeed, it was galaxy studies that first gave a hint that there is more to the Universe than meets the eye, although for many years most astronomers were reluctant to take that hint.

Way back in the 1930s, only a decade or so after astronomers recognized that some of the fuzzy patches of light observed through their telescopes were other galaxies beyond the Milky Way, the Swiss astronomer Fritz Zwicky noticed a peculiar thing about clusters of galaxies. In many cases, the galaxies in these clusters were moving too fast for the clusters to be held together by the gravitational pull of all the bright stars in all the galaxies in the cluster. If the observations were right, the clusters could not be stable but ought to be evaporating on a rather rapid timescale, by astronomical standards. At the time, both the idea of external galaxies and the use of Doppler shifts (*not* the cosmological redshift) to measure the speeds with which such galaxies were moving were new, and few people took Zwicky's results at face value. But if you did accept them, they implied that in order for clusters of galaxies to be stable (or "gravitationally bound") there must be several hundred times more gravitating matter in large clusters of galaxies than there was in the form of bright stars. Zwicky referred to this unseen material as "dark (cold) matter [dunkle (kalte) materie]." Even if you did take the results seriously, at that time there was no reason to think there couldn't be that much dark baryonic matter around in the form of cold clouds of gas or very faint stars, so it wasn't much of a worry. Even though the possibility of dark matter was acknowledged nearly seventy years ago, it only started to be a cause for concern after the development of an understanding of Big Bang nucleosynthesis in the 1960s set limits on the amount of baryonic material around. Then, nearly forty years after Zwicky's pioneering work, in the 1970s another variation on the dark matter theme cropped up.

At that time, several researchers were studying the way disk galaxies, like our Milky Way, rotate. As their name implies, disk galaxies are flattened star systems, rotating disks with a central bulge, having roughly the proportions of a fried egg but with diameters typically of a hundred thousand light years, and containing hundreds of billions of individual stars. The whole system wheels around like a leisurely Catherine Wheel, taking a couple of hundred million years for a star like the Sun (two-thirds of the way out from the center of our own Galaxy) to complete one orbit around the center. When we see such galaxies edge on, it is possible to measure how fast they are rotating using the Doppler effect. One side of the disk will be coming toward us, so its light will show a blueshift, and the other side will be moving away from us, so its light will show a redshift. The size of the shift reveals the speed with which the disk is moving. By the 1970s, the technology was good enough that in many cases the speed with which different parts of such a galaxy were moving could be measured at different distances out from the center. The results came as a big surprise.

If all the mass of a disk galaxy were distributed in the same way as the bright stars, the stars farther out from the center would be moving more slowly in their orbits because they are farther away from the concentration of mass in the central bulge, or nucleus. In the same way, in our solar system the outer planets such as Jupiter and Saturn move more slowly in their orbits than the inner planets such as Venus and Earth, because they are farther away from the concentration of mass in the Sun. But in virtually every case the observations of disk galaxies show that except in the innermost regions of the galaxy being studied, the orbital velocity is the same for stars way out on the edge of the disk as it is for stars orbiting close to the central bulge, and for all points in between. The only viable explanation for this is that disk galaxies are embedded in huge haloes of dark material, containing at least ten times as much matter as in the disk itself, which holds the disk in its gravitational grip. This directly shows the presence of dark matter associated with individual galaxies, while the cluster studies pioneered by Zwicky show that there must also be additional dark matter in the gaps between galaxies.

In the early years of the twenty-first century, astronomers found more direct evidence of the existence of these clouds of dark matter between the galaxies. Remember that only about a fifth of the number of baryons that were produced in the Big Bang can be seen today in the form of bright stars and galaxies. The rest must be around somewhere, in clouds of gas between the stars and galaxies or in faint stars. For a long time nobody knew exactly where, but it seemed a natural guess that this baryonic dark matter was invisible because it was cold. This guess turns out to have been exactly wrong. It is invisible because it is hot.

The "dark" baryonic matter has been located by satellite observations made in the ultraviolet part of the spectrum, which is invisible to our eyes. They have revealed that our Galaxy and its neighbors (a kind of mini-cluster called the Local Group of galaxies) is embedded in a great fog of hot intergalactic gas, in the familiar form of hydrogen and helium. Although it is very tenuous by terrestrial standards, this gas is very hot, in the sense that the particles in it move very quickly and radiate at short wavelengths, beyond the blue end of the visible spectrum, in the ultraviolet. If we had ultraviolet eyes, we would see it as a haze of hot bright matter covering the entire sky, with a temperature of about 10–20 million K (1–2 keV). There is a mass equivalent to about a trillion times the mass of the Sun in this hot stuff surrounding the Local Group, roughly four times as much as the mass in the bright stuff in galaxies and neatly matching our understanding of Big Bang nucleosynthesis. But this still leaves scope for plenty of cold dark matter. Like the evidence from the speed with which galaxies move which shows that clusters are held together by dark matter, such a hot cloud of gas can be held in place only by cold dark matter. The hot gas is embedded in the cold dark matter and provides us with a tracer of the dark matter, like fairy lights providing a trace of the outline of a Christmas tree.

Similar clouds of hot gas have now been identified surrounding other galaxies in other clusters. The evidence suggests that the gas is kept hot by blasts of energetic material thrown out by the active centers of some galaxies, which are detected as energetic radio sources and are probably associated with supermassive black holes. The gas may also have been made hot originally by shock waves produced by collisions between clouds of gas in the early Universe, not long after recombination.

One way and another, by the time the computer simulations of the growth of structure in the entire visible Universe showed the need for cold dark matter, it was no longer really a surprise. But there is still one major puzzle to resolve.

Our understanding of primordial nucleosynthesis tells us that about 4 percent of the mass needed to make the Universe flat is present in the form of baryons (with less than half of 1 percent in the form of neutrinos), and we know from observations that about a fifth of this (less than 1 percent of the flatness requirement) is in the form of bright stuff. The comparison between the pattern made by the distribution of bright stuff in the Universe today and the patterns made in computer simulations tells us that the total amount of matter in the Universe is about 30 percent of the mass needed for flatness — in other words, about 26 percent of the flatness mass, between six and seven times as much mass as there is in the form of baryons, is in the form of cold dark matter. Any more, and the pattern made by the bright stuff would be more clumpy; any less, and it would be less clumpy.

But a combination of the computer simulations and the studies of the cosmic microwave background radiation tells us that the Universe is flat. One way to understand this is to appreciate that the rate a which the Universe is expanding also affects its clumpiness today. If the Universe were open, it would expand faster, matter would get stretched thinner more quickly, and there would not have been time for structures as large as the ones we see to have formed since the Big Bang; on the other hand, if the Universe were closed, it would expand more slowly, matter would clump together more easily, and the Universe would be more lumpy than we actually see it to be.* This is the puzzle — what is it that makes the Universe flat, if only 30 percent of the flatness mass is in the form of matter? The resolution to that puzzle will be explained in the next chapter; but first, here is a summary of how we think structure evolved in the Universe as it expanded away from the time of recombination.

We know that concentrations of baryonic matter could not begin to grow until after recombination, because the interactions between charged particles and the still hot photons of the background radia-

---

*In case you are wondering, you can't trade off a change in the expansion rate and a change in the amount of dark matter to get a different "fix" for this problem; the rather delicate balancing act between all the parameters involved is trickier than the simplistic examples used here might suggest, and it only works out for this thirty/seventy split.

tion would have prevented the collapse before then. But we also know that the dark matter must have already been concentrated in clumps by that time because of the speed with which baryonic material did fall into the gravitational potholes once it became locked up in electrically neutral atoms. Indeed, one of the surprises of the early twenty-first century was that as the technology improved and observers were able to look back farther in time to higher and higher redshifts, they kept finding protogalaxies and clouds of hot hydrogen gas at every stage back in time. The best explanation for this is, as we shall see, that black holes formed very early in the life of the Universe and acted as the seeds on which galaxies grew. Some calculations suggest that the primordial black hole seeds formed out of density fluctuations in the Universe around the time of primordial nucleosynthesis. This is still only a hypothesis — an example of what we *think* we know — but it is the best explanation anyone has come up with yet.

An analysis of the Hubble Ultra Deep Field made in 2004 shows that at a redshift of 6, about 900 million years after the Big Bang, the Universe contained many small, faint objects called dwarf galaxies. It was the ultraviolet light from these galaxies that completed the process of reionization. At slightly higher redshifts, however, about 700 million years after the Big Bang, there were noticeably fewer of these dwarf galaxies, which suggests that we are seeing the epoch when the formation of these small galaxies peaked.

It certainly didn't take the dwarf galaxies long to merge with one another and form larger galaxies. In another study reported in 2004, astronomers analyzed the light from galaxies that were around when the Universe was between three and six billion years old (between eight and eleven billion years ago), which are visible from ground-based telescopes. The striking feature of the galaxies studied in this survey is that they are "mature" systems which look very much like the galaxies we see in the nearby Universe today. They have already gone through their major early phase of merging and star formation and settled down into a relatively quiet state. Even a cluster of galaxies, certainly a mature system in this sense, has been detected at a distance of nine billion light years, five billion years after the Big Bang. In a stunning example of the power of modern technology, the cluster was first identified in an X-ray image obtained by the European satellite XMM-Newton, and it was revealed by just 280 photons gathered by the telescope in a 12.5-hour exposure of a tiny patch of sky. When ground-based optical telescopes were directed at the spot, they found twelve large galaxies at the heart of what is probably a cluster of hundreds of smaller galaxies (too faint to be seen from Earth) held together by gravity. The discovery was announced in the spring of 2005, and the technique could be revealing more clusters at distances like this by the time you read these words.

As well as disk galaxies like our Milky Way (which are also sometimes called spiral galaxies, though this is not really a good name since by no means all disk galaxies show a spiral pattern), the present-day Universe contains many elliptical galaxies (which range in shape from spherical to the "oval" of an American football, and come in many different sizes) and some remaining small, irregular dwarf galaxies. Observations show that the whole range of types was already established in the Universe by the time it was a third of its present age, and these galaxies were already gathering into well-defined clusters by then. All of this is evidence that seriously large concentrations of dark matter were present to act as the seeds for galaxy formation right from the time of recombination — but we have frustratingly little direct observational evidence from redshifts greater than 7.

The planned successor to the Hubble Space Telescope, the James

Webb Space Telescope (JWST),* should be able to look back even farther, to redshifts of 20; but until the JWST is launched, which won't be before 2011, astronomers have to rely on chance alignments of galaxies along the line of sight with objects at very high redshifts to get a glimpse of what was going on that close to the Big Bang.†

When this happens, the gravity of the intervening galaxy (or of a whole cluster of galaxies) can bend the light from the more remote object and concentrate it like a huge magnifying glass. Such a gravitational lens acts as a kind of natural telescope far more powerful than any artificial telescope; the necessary alignments are rare and usually produce a somewhat distorted image of the distant object, but even a few distorted images are better than no images at all.

The farthest known individual galaxy (at the time of writing, summer of 2005) was found in this way. A long-exposure image of a nearby cluster of galaxies called Abell 2218, taken with the HST, revealed the distorted image of a more distant galaxy superimposed on the image of the cluster. Analysis of the light from the object implies that it has a redshift close to 7, corresponding to a look-back time of 13 billion years, and that we are seeing it by light which left when the Universe was only 5 or 6 percent of its present age. It is hard to estimate the size of this protogalaxy because of the way the image is distorted, but it seems to be only about 2,000 light years across,

---

*It's a sign of the times that whereas the HST was named after a pioneering astronomer, the JWST is named after a NASA administrator!

†Because of the way the equations work, although for nearby objects (with redshifts less than about 1) doubling the redshift implies doubling the distance, close to the Big Bang the redshift goes off the scale, so that the Big Bang itself is at a redshift of infinity. So, for example, a redshift of 7 corresponds to a distance of about 13 billion light years, but redshift 20 corresponds to "only" about 13.5 billion light years, not to 39 billion light years. The cosmic background radiation originates from a redshift of about a thousand.

although it shines relatively brightly in the ultraviolet part of the spectrum. This is a hint that stars are (or were!) forming actively in the young galaxy, because young stars are usually hot and produce a lot of blue and ultraviolet light. This neatly fits in with estimates of the time of reionization, since reionization is thought to have been caused by ultraviolet radiation from young galaxies. And that in turn suggests that this unspectacular object really might be one of the very first galaxies to have formed. In another study, also taking advantage of a natural gravitational lens, astronomers identified an even smaller object, a cluster of stars rather than a cluster of galaxies, which also lies just over 13 billion light years away. Such clusters are spherical collections of stars held together by gravity that contain about a million individual stars; they are common components of galaxies like the Milky Way. All of this is powerful evidence that the large galaxies seen in the Universe today formed from the accumulation and merger of smaller units that formed first — the "bottom up" method of building structure in the Universe, which is still continuing today.

There is just one more ingredient to add to the mix — energetic objects known as quasars, which are thought to be powered by supermassive black holes, containing as much mass as millions of stars like the Sun, although a lot of this mass might originally have been dark matter. The name *quasar* is a contraction of *quasi-stellar,* and it is apposite because on short-exposure images of the sky quasars look like stars; but they are not. Longer-exposure images reveal that quasars are enormously bright objects at the hearts of some galaxies, shining so brightly that it is hard to see the surrounding stars of the galaxy itself, just as it would be hard to see the glow from a few candles in the glare of a searchlight alongside them. The energy which makes them shine so brightly is thought to be released as they swallow matter from the inner region of the surrounding galaxy, and it is also thought that all large galaxies (including our own Milky Way) have a black hole at their center, although in many cases this is no longer active because it has swallowed up all the material nearby.

This is so far from the popular image of a black hole as a collapsed star with no more than a few times the mass of the Sun that it is worth giving a little more detail about the nature of these beasts. A black hole is a collection of matter which has a gravitational pull so strong that nothing, not even light, can escape from it. It is true that one way to make a black hole would be to take a few solar masses of stuff (any stuff) and crush it into a space a few kilometers across. This happens to some stars at the ends of their lives, and many such black holes — called, logically enough, "stellar mass" black holes — have now been identified in our Galaxy. But the same trick could also be achieved by making a very large object with a rather low overall density. A few million stars like the Sun packed into a sphere with the same radius as our Solar System out to the orbit of Neptune (like a rather large bag of marbles) would only have the same density as the oceans of the Earth, but it would be a black hole. Nothing could ever escape from it. This is the kind of supermassive black hole, only as big as the Solar System, that resides at the hearts of galaxies and powers quasars.

Such a supermassive black hole has a strong gravitational pull and attracts matter toward itself. But it is quite difficult for the large amount of matter attracted in this way to get into the black hole, because it has such a small surface area. So the infalling matter piles up in a swirling disk around the black hole, gradually funneling its way in. The material in the disk moves rapidly because it is under the influence of such a strong gravitational pull, and it swirls around, getting hot as atoms collide with one another. This converts gravitational energy into heat, light, radio, and X-rays, all of which makes the quasar shine brightly as long as it has a source of matter to feed off.

The process is so efficient that up to half of the rest-mass energy of the infalling matter gets converted into radiant energy, in line with Einstein's famous equation, and the black hole only has to swallow about one solar mass of material each year to keep the quasar shining. But eventually its fuel supply will run out, and there are few quasars still active in our neighborhood of the Universe today.

Such black holes must have been important seeds for the growth of structure early in the history of the Universe, and the fact that structure grew as quickly as it did suggests that black holes were present from the earliest moments after recombination, and may have been produced by processes which are not yet fully understood involving the dark matter, before the baryonic stuff cooled sufficiently to collapse into stars and galaxies. The observations of quasars, seen at redshifts all the way back to 6.5, show that black holes as large as any in the Universe today (perhaps containing a billion times the mass of the Sun) were already present within a billion years of the Big Bang, when the Universe was less than a tenth of its present age. But that is as far back as we can yet push the observations of galaxies and quasars. It's time to take on board the best, but admittedly still speculative, explanation of how structure emerged in the Universe after the Big Bang.

Until a redshift of about 100, 20 million years after the Big Bang, the Universe was still nearly smooth, but then structure began to grow rapidly. The most likely candidates for cold dark matter (discussed in more detail in the next chapter) are particles which each have a mass about a hundred times the mass of a single proton. The observed pattern of structure in the Universe shows that individual CDM particles must have masses less than 500 times the mass of a proton — less than 500 GeV — while accelerator experiments have ruled out any mass less than about 40 GeV. They only interact with baryons through gravity or if they happen to bump into a baryon and give it a knock. At a redshift of 100, the baryons are still far too hot to collapse and form compact objects, but computer simulations show that the CDM particles will quickly collapse under the pull of gravity, starting from the kind of ripples seen in the background radiation, until by a redshift of between 25 and 50 they have formed spherical clouds of dark matter as massive as the Earth but as big as our Solar System, with most of the mass of each cloud concentrated in its center. The clouds themselves then cluster together as a result of their mutual gravitational pull, resisting the expansion of the Universe to form clusters of clouds, and clusters of clusters, in which, once again, most of the mass is concentrated in the center.

By the time the baryons had cooled sufficiently to collapse, the dark matter potholes were well developed — with black holes at the centers of the larger aggregations of these clouds — so that baryonic stuff streamed in toward the concentrations of dark matter, forming stars and galaxies as it did so. This scenario neatly explains the presence of bright objects with masses ranging from a few million times the mass of our Sun (like the spherical clusters of stars) to hundreds of billions of times the mass of our Sun (galaxies like the Milky Way) and structures tens of thousands of times bigger still (superclusters of galaxies). The calculations also suggest that huge numbers of the original Earth-mass dark matter clouds should have survived to the present day, with as many as a million billion of them (10^{15}) in the spherical halo of dark matter surrounding our own Galaxy. Interactions between the dark matter particles within those clouds should produce showers of gamma rays that are too feeble to be detected on Earth but which might possibly be detected by the next generation of satellite experiments, before about 2012.

Calculations show that even at a redshift of 6.5, scarcely more than a billion years after the Big Bang, there would have been enough time since recombination for this process of bottom-up growth to have produced black holes with masses of a billion or more solar masses, embedded in dark matter halos containing about a thousand billion solar masses of material, with baryonic matter falling onto the black hole to provide the energy to power the quasar while stars formed out of the clouds of baryons farther out from the center of the object. But the growth would only have been rapid enough to form structures as big as the Milky Way in the time available if the central black hole had a mass of at least a million solar masses; happily for the theorists, the mass of the black hole at the heart of the Milky Way itself turns out to be about three million times the mass of our Sun.

The variety of galaxies seen in the Universe today is largely a result of mergers. Collisions and interactions between galaxies can be seen in many clusters, and without going into detail it is easy to see in general terms how a large collapsing cloud of material will naturally settle down into a disk, like our own Galaxy, as it rotates. Larger galaxies will gobble up small ones if they come too close, but collisions between disk galaxies, perhaps including the merging of the black holes in their centers, can produce bursts of star formation and strip away the material of the disks, forming ellipticals; as an elliptical formed in this way settles down after such an event it may "grow" a new disk as more baryonic material settles down around what is now the central bulge. Small irregular galaxies are simply left over from the early days of the Universe — and the observations show, as you would expect, that there were far more small galaxies when the Universe was young compared with today. They have been swallowed up in mergers to make the large galaxies we see today.

The largest galaxies in the Universe are all ellipticals, and some of them are very large indeed. About 60 percent of all galaxies are ellipticals, but the largest contain as much as a trillion (10^{12}) times the mass of the Sun, so an even higher proportion of the baryonic mass is in ellipticals. In fact, three-quarters of the mass in stars in the entire Universe is in the form of these giant ellipticals, which are seen back to redshifts of 1.5 or more. Beyond that, they are too faint to be detected directly. From the colors of their stars, however, it is clear that they were already old by then, and that some formed at redshifts of 4 or 5 — or, at least, the components that merged to make them had formed by redshifts of 4 or 5.

Our own Galaxy seems to be a little more than ten billion years old; but our Sun and Solar System are less than half this age, about 4.5 billion years old. Clearly, star formation continued long after the first galaxies formed, and, indeed, we can see stars still forming in our Galaxy today. This is convenient, because it helps us understand where stars came from. In particular, it helps us understand how the Solar System we live in began. But before we leave the Universe at large and focus on this topic of particular interest to human beings, there is one piece of leftover business to resolve. As we mentioned, the combination of observations, computer simulations, and theory tells us that the total amount of matter in the Universe is 30 percent of the amount needed to make the Universe flat. But the same combination tells us that the Universe *is* flat! If it were open, it would fly apart too quickly for galaxies like the Milky Way to have formed in the time available, and we would not be here to wonder how things began. So what happened to the other 70 percent? What is it that holds the Universe together?

6

What Is It That Holds
the Universe Together?

When cosmologists first realized that there is more to the Universe than meets the eye, and more even than can be explained in terms of dark baryonic matter, their natural first assumption was that there must be nonbaryonic stuff out there, floating around in the form of exotic (compared with terrestrial stuff) particles or lumps in the space between the visible stars and galaxies. This assumption has been borne out by subsequent observations of the Universe, which strongly suggest that exotic dark matter of this kind does exist. But those same observations tell us that even adding the maximum possible amount of these exotic particles to all the baryons in the Universe does not provide enough matter to make the Universe flat. This implies that in addition, there is still a third component to the Universe, which has become known as dark energy. As we have seen, all the matter in the Universe, baryons and exotic particles combined, adds up to only 30 percent of the density needed for flatness. Nevertheless, the exotic dark matter still makes a significant contribution to holding the Universe together (six or seven times as important a contribution as baryonic matter), so we will look at it first, before coming to grips with dark energy.

Apart from the tiny fraction of the total mass provided by neutrinos, all this exotic dark matter must be cold, in the sense that the particles move much more slowly than the speed of light. It is generically referred to as cold dark matter (CDM), although some astronomers with a fondness for acronyms use the term WIMPs, from Weakly Interacting Massive Particles. (*Massive* here just means that they have *some* mass, not that they are particularly heavy.) Both terms refer to the same thing, and although it would be conveniently simple if all the CDM were made up of the same kind of stuff, there is nothing in the observations to tell us this. For all we know there could be several different kinds of WIMPs, as long as their total mass added up to about 26 percent of the amount required for flatness. In fact, though, the particle physicists have only been able to come up with two good candidates for CDM particles. This may be a result of their limited imagination, but it helps keep the picture from becoming too complicated. CDM could be entirely composed of one or other of these types of particle, or any mixture of the two, provided the total mass added up to that critical 26 percent. As you can see, we are already well into the area of what we *think* we know, rather than what we think we *know* — and it will get worse when we move on to dark energy.

The first candidate for CDM is called the axion. The name is appropriate, because its existence (if it does exist) is related to a property of particles called spin, which can be thought of in terms of little spheres rotating on their axes, although, as with all analogies in the quantum world, this picture only represents part of the truth. It also happens that "Axion" was the name of a laundry detergent used in the United States from which the physicists involved took the name; it isn't just astronomers who sometimes take a childish delight in choosing names for newly discovered entities.

The need for the axion first emerged at the end of the 1970s, when particle theorists were struggling to come to grips with a curious implication of quantum chromodynamics (QCD) which suggested that there might be some processes of particle decay which violated the principle of time symmetry—in other words, the interactions involved would only "work" for one direction of time. This was alarming because one of the most cherished tenets of theoretical physics is that all such interactions work equally well "forward" or "backward" in time, like a video of a collision between two pool balls, which makes perfect sense whichever way you run the tape. In order to restore time symmetry to the interaction,* the theorists had to invoke a new field; like all fields in the quantum world this new field had to have a particle—in this case, the axion—associated with it.

Early versions of this model suggested that the axion ought to have a relatively large mass and therefore might be detected in accelerator experiments within a few years. But in the 1980s, just as the failure to find the axion was beginning to be embarrassing, QCD was incorporated into the grand unified theories (GUTs), and in this context the model required a much smaller mass for the axion, making it far too light to be detected directly. This was dubbed the "invisible" axion, and dismissed as something of a joke by many physicists. What was the point of inventing a particle too light to be seen in order to explain away a tiny bit of symmetry-breaking that might in any case be real?

---

*And I have to confess that I would not personally be too downhearted if time symmetry were broken in this subtle way.

But when cosmologists began to appreciate the need for dark matter, the axion was a candidate ready and waiting to be fitted to the role. It also turns out that the axion arises naturally in the context of string theory.

The other key property of the axion, predicted by all variations on the theme, is that in spite of its tiny mass, the way it is manufactured naturally in particle interactions sets it off traveling at low speeds, compared with the speed of light. If the ideas are correct, huge numbers of axions would have been produced in the Big Bang, at about the time quarks were condensing out to make protons and neutrons, but they would indeed be *cold* dark matter particles. So axions, rather than streaming through space and smoothing out incipient structure in the early Universe like heavy neutrinos, would clump together under the influence of their own gravity, making the potholes into which baryonic matter would fall. Together, they *could* provide all the cold dark matter.

The best way to get a handle on the masses of *individual* axions turns out to come from astrophysics, not particle physics. Because axions interact only weakly with baryonic matter, just like neutrinos they would be able to stream out from the central cores of stars into space virtually unobstructed, carrying energy away with them and cooling the stellar cores. This cooling is more effective the heavier the axions are (it is almost negligible for neutrinos), and if each axion had a mass greater than 0.01 eV this would affect the appearance of stars and the way some old stars explode as supernovas in observable ways. Since these effects are not observed in real stars, the mass of the axion must be less (probably at least a factor of ten less) than 0.01 eV, or 0.002 millionths of the mass of an electron. The theoretical models suggest that the mass could be lower still, below 0.0001 eV. It's hardly surprising that they have not been found in accelerator experiments here on Earth!

As the tone of my comments may have suggested, I am not a fan of the axion, although it has to be admitted that it *could* exist. One reason for my lack of enthusiasm is that even if invisible axions do exist, it will be almost impossible to detect them. Indeed, I know of only one serious proposal to search for axions, and even that seems to be a triumph of hope over expectation.

The hope is that axions might be detected as a result of their extremely rare interactions with electromagnetic fields. But the futility of this hope as far as any realistic prospect of detecting axions in the foreseeable future goes is highlighted when you realize that neutrinos, reluctant though they are to interact with other stuff, are ten billion times more likely to interact than axions are. Nevertheless, theory suggests that very, very occasionally an axion will interact with an electromagnetic field to produce a photon, with a wavelength that depends on the mass of the axion. And since there are (if the axion idea is correct) vast numbers of axions around, it is possible that one day we might have the technology to detect those photons. But for every ten billion axions around, there is as much chance of a single detectable interaction as if you were trying to detect a single neutrino.

There is only one remote chance of this kind of interaction being detected in the not too distant future. Because all the photons would have nearly the same wavelength — apart from a spread caused by the motion of the individual axions through space — they would, like the photons in a laser beam (but much, much fainter), add up to produce a detectable pulse of noise (in this case, radio noise), a spike in the electromagnetic spectrum sharply located around one wavelength. The detector would work like this. First, you need a metal box (what physicists call a cavity) just the right size for photons with the right wavelength to form standing waves inside the cavity — this is the same principle as tuning organ pipes to produce standing sound waves with a particular note, but applied to electromagnetic waves. The cavity has to be shielded from outside interference, cooled in a bath of liquid helium close to absolute zero (-273 °C), and completely empty of ordinary matter, so that it contains only the inevitable neutrinos (which, happily, would not produce the same kind of signal as axions) and whatever cold dark matter particles there are in the Universe, including, if they exist, axions. The box then has to be filled with the strongest magnetic field that can be produced, and sensitive radio detectors tuned in to listen for the axion signal.

For a cavity in the form of a hollow cube 1 meter on each side, filled with the strongest magnetic field that we could force into the box, the power output of the predicted axion signal would be just one-millionth of a billionth of a billionth (10^{-24}) of a watt. Physicist Lawrence Krauss has put this in perspective by calculating that an axion detector the size of the Sun would produce the same power output as a 60 watt lightbulb. So it's hardly surprising that most rational people do not expect to see proof that axions exist and would prefer to have a dark matter candidate that we might at least be able to detect.

Fortunately for those of us who think this way, there is a much stronger candidate for the role of CDM, which emerges naturally (indeed, inevitably) from the idea of supersymmetry, and which will definitely be detected in the near future if it does exist. As we saw in Chapter 2, supersymmetry (SUSY) implies the existence of a variety of supersymmetric partners, a counterpart to each and every kind of known particle, though only the lightest supersymmetric partner (LSP) is predicted to be stable. This immediately suggests that the LSP is a good candidate for the cold dark matter particle needed by the cosmologists.

One minor inconvenience is that our present understanding of supersymmetry does not actually tell us what the LSP is. It might be the photino (the SUSY counterpart to the photon), or the gravitino (the SUSY counterpart to the graviton), or one of the other SUSY particles. According to some versions of the theory,* it might even be a mixture of two or more different kinds of particle, in the same way that neutrinos travel through space as a mixture of the three kinds of neutrino detected in experiments. But one thing we know for sure is that the LSP has no electric charge, because if it did it would be easy to detect — it would, indeed, have been detected long before astronomers appreciated the need for CDM. So in order to leave all the options open, the LSP is often referred to simply as the neutralino (in other words, the "little neutral SUSY particle"). The name neutralino is not the name of any specific particle, but a generic name to cover all the options for the LSP.

Although the original SUSY theory said that the mass of the neutralino could be as little as a few GeV (remember that 1 GeV is roughly the mass of a proton, or a hydrogen atom), the fact that such particles have not yet been produced in accelerator experiments tells us that their mass must actually be above 50 GeV. At the other end of the scale, we can set an extreme upper limit on the mass of the neutralino from cosmology. As we saw in Chapter 5, the presence of additional particles (such as more varieties of neutrino) in the Big Bang makes the Universe expand *faster*, by pushing it outward more vigorously. Neutralinos would have the same effect, and if the mass of each neutralino were bigger than about 3,000 GeV the Universe

---

*I'm a little unhappy with my own use of the term *theory* here and in other places in this book. If I were writing for an academic readership I would be more careful to distinguish theories (which are soundly rooted in experiment and observation) from models and hypotheses, which are more speculative. But *theory* also has a wider usage among nonscientists, and that is the way I use it here.

would have expanded so rapidly that we would not be around to study it. This is not a very restrictive limit, but some versions of SUSY (which the theorists like to think of as improved versions) suggest an upper limit of about one-tenth of this, around 300 GeV, or three hundred times the mass of a hydrogen atom. The best bet at present seems to be that the neutralino has a mass somewhere in the range from about 100 GeV to 300 GeV (roughly corresponding to the masses of individual atoms of the heaviest naturally occurring elements on Earth; a uranium nucleus, for example, has a mass close to 235 GeV). Happily, this is just the range that can be probed by the next generation of accelerator experiments, and for which direct detection of dark matter particles is possible, promoting this package of ideas from the realm of speculation to the realm of real science.

If there is seven times as much matter in the form of CDM as in the form of baryons, and if, say, each CDM particle had a mass of 140 GeV, then, since the average mass of a baryon is about 1 GeV, there would be one neutralino for every twenty baryons in the Universe. If they were spread out evenly through the Universe, that would imply just one neutralino in every 4 cubic meters of space — but, as we have seen, they must actually be clumped together rather like the bright stuff of the Universe, so there should be rather more than that passing through the Earth and our laboratories (and, indeed, our bodies), and capable of being detected.

Apart from making neutralinos out of energy in particle collisions at the LHC, which is certainly a possibility if they have masses at the lower end of the predicted range, two ways to detect neutralinos in the lab are now being pursued. Both depend on the fact that just about the only time baryonic matter "notices" the existence of neutralinos (except through gravity) is when a neutralino physically collides with the nucleus of an atom and bounces off it. For atoms (nuclei) with roughly the same mass as the neutralinos, this scattering, as it is called, is rather like a collision between two pool balls, with the nucleus that has been struck recoiling while the incoming neutralino bounces off in a new direction. Such an event is most likely to have a noticeable effect if the nucleus struck in this way is in a very well-ordered solid, part of a neat lattice arrangement of atoms all of the same kind in a regular pattern. In order to minimize the natural jiggling about of the atoms in such a lattice, it has to be cooled to a fraction of a degree above the absolute zero of temperature, $-273\ °C$; and in order to minimize interference from other interactions (for example, from cosmic rays) it has to be shielded from the outside world. These are not easy requirements to meet. But once those conditions *are* met, it starts to become possible to contemplate detecting neutralinos.

If such a collision — essentially the same as a collision between pool balls — occurred between a neutralino and a nucleus in a supercooled, shielded lattice, a block of a substance such as silicon or germanium, there are two effects that might, in principle, be detected. One is that the recoiling atom could shake up its neighbors to produce a tiny ripple spreading through the block, a very weak sound wave. If the block of material were covered in a film of superconducting material, the effect of the sound waves on the superconductor could be measured. The lattice of atoms is held together by electromagnetic forces, as if each atom were joined to its immediate neighbors by tiny springs, and you can picture this as all the little springs bouncing in and out in a regular way as the sound wave passes through the lattice. The technique has actually been tested by bombarding suitably prepared samples of silicon with "ordinary" radiation, and it works; but as yet it has not been able to detect neutralinos.

The other way to detect neutralino collisions with ordinary nuclei in blocks of supercooled material is simply by the rise in temperature produced in the sample as the energy of motion of the neutralino first makes the "target" nucleus recoil and then sets the nearby atoms jostling about in an irregular, disordered way as the energy is shared among them through the vibrating "springs." The energy deposited in this way would only be a few keV, so the rise in temperature resulting from a neutralino impact on even a small block of silicon would only be a few thousandths of a degree—but if the target sample is already at a temperature of only a few thousandths of a degree above absolute zero, that could mean doubling its temperature! Again, the techniques required to measure such temperature changes have been tested, and they work. This time, however, there have already been claims (unfortunately, unconfirmed) of the detection of a dark matter "signal."

The experiment that seemed, for a time, to have detected dark matter particles is called DAMA (from DArk MAtter); it is based in a mine at Grand Sasso, under the Apennine Mountains of Italy, where it is well shielded from everything outside. The detector, built around a crystal of sodium iodide, has been running for several years, and seemed, according to data released at the beginning of the twenty-first century, to show a seasonal fluctuation. One possible explanation for this is that as the Earth moves around the Sun, on one side of the Sun it is moving head-on into the halo of neutralinos rotating with our Galaxy, and on the other side of the Sun it is moving with the neutralinos. Just as in a car crash, head-on collisions will deposit more energy in the crystal, and "overtaking" collisions will deposit less, so such a seasonal effect might well result from the presence of neutralinos. The DAMA team even claimed to have a handle on the mass of the neutralino, somewhere between 45 GeV and 75 GeV. Unfortunately (or perhaps fortunately, since the mass suggested is rather low), other experiments which ought to be just as sensitive as DAMA — including one dubbed the Cold Dark Matter Search (CDMS) which uses germanium and silicon detectors at its base down a mine in Minnesota — have found no such effect.

In my view, this kind of detector is the most likely way we will find cold dark matter particles, and it could happen very soon. My favorite experiment of this kind is based down a mine at Boulby, in Yorkshire, and since there is such a good chance of identifying neutralinos soon, it seems worth describing this experiment in a little more detail.

Because the dark matter candidates (neutralinos) are weakly interacting, assuming they exist there will be less than one collision per day between a neutralino and one of the atomic nuclei in a lump of matter weighing 10 kilos. Although there are far fewer baryonic cosmic rays around than there are neutralinos, because these interact much more readily with lumps of everyday matter they will produce far more collisions each day, which is why the Boulby experiment is set up in salt caverns 1.1 kilometers below ground at the bottom of Europe's deepest mine. The layer of rock above these caverns stops all but one in a million of the cosmic rays; but of the billion or so WIMPs per second passing through a human being walking on the surface above the mine, just three will collide with nuclei in the rock on the way down to the caverns, and the collisions will only slow them down, not stop them.

Even this "filtering" is not enough to reduce the level of background noise (the equivalent of static on an AM radio) from cosmic ray interactions to the same level as the frequency of dark matter events in the detectors, and an additional allowance has to be made for the natural radioactivity of the rocks surrounding the cavern. A lot of this radiation, though, can be absorbed in shielding material surrounding the detectors — mundane materials such as lead, copper, wax, and polythene—or even by immersing the detector in a tank containing 200 tonnes of pure water. That much water (for screening alone) occupies a volume of 200,000 liters, about a tenth of the volume of an Olympic swimming pool and half the size of the neutrino detector mentioned in Chapter 5.

Even after all those precautions have been taken, there is still background noise in the system to contend with, so the last stage is to use detectors and statistical techniques which can distinguish between the different kinds of events produced by the background radiation and by the recoil of atoms (strictly speaking, nuclei) from pool-ball type collisions with neutralinos. This is where the sound-wave and temperature-rise techniques we described earlier come in.

In spite of all the difficulties, the biggest concern of the Boulby team is not that their detectors will fail to detect neutralinos but that the Large Hadron Collider might beat them to it. Depending on their exact mass, neutralinos could be produced in the collisions between beams of protons at CERN before the end of the first decade of the twenty-first century; but it is harder to make heavier particles, and if neutralino masses are as high as the favored SUSY models suggest, the detection is likely to be made first at the Boulby mine, or at one of the similar experiments now running around the world.

Even if the Boulby experiment, or one of the other dark matter experiments, does find the elusive neutralino, we will still only know the composition of 30 percent of the stuff that makes the Universe flat. By the middle of the 1990s, it was already becoming clear from a comparison of simulations of galaxy clustering with maps of the real Universe that there could not be more than about 30 percent of the flatness density in the form of matter (baryonic or otherwise) clumped together like galaxies, and it was also a clear requirement of inflation that the Universe must be flat. In any case, as we discussed in Chapter

3, many cosmologists had long felt that the Universe must be precisely flat, because any deviation from flatness would have grown exponentially fast as the Universe expanded away from the Big Bang. This was certainly an argument that had impressed me since my student days in the 1960s. Anyone who shared that view automatically had to accept that the other 70 percent of the Universe must be in the form of a completely uniform, unclumped stuff that had no major effect on galaxy formation (because it made no "potholes") but which did affect the structure of space-time. Happily for those cosmologists, something that exactly fitted the bill had been found by Albert Einstein back in 1917 — although he had discovered it for the wrong reason and later abandoned it entirely.

Einstein completed his general theory of relativity in 1916. It is a theory that describes the interaction between matter, space, and time operating through gravity. The first thing he did with his theory, once it was complete, was apply it to provide a mathematical description of the biggest thing involving matter, space, and time that there is — the Universe. (There is a sense in which the general theory *only* applies perfectly to the description of a complete universe, with no edges — boundary conditions — to worry about, so this was a natural thing for him to do.) This was in 1917, when many astronomers still thought that our Milky Way Galaxy was the entire Universe, and the fuzzy patches of light then known as nebulae had yet to be identified as other galaxies beyond the Milky Way. The received wisdom was that the Universe was essentially static and unchanging — individual stars might be born, run through their life cycles and die (like individual trees in a forest), but the "forest" of the Milky Way would always present much the same overall appearance.

Einstein immediately ran into a snag. His equations, the equations of the general theory of relativity, did not, in their simplest form, allow for the possibility of a static universe. The equations contained within themselves a description of expanding universes, in which gravity was acting to slow the expansion down, and a description of collapsing universes, with gravity acting to speed the collapse up. But there was no description of a universe balancing on the knife-edge between these two scenarios. The only way to have such a universe would be if there were some effect opposing the pull of gravity that would in effect cancel gravity out and allow everything to hang in place, balanced on the knife-edge between expansion and collapse. But adding the smallest possible complication to the equations, a number which Einstein called the cosmological constant, made this balancing act possible. (The cosmological constant really is the simplest possible addition to the equations, an example of what is known as a "constant of integration.") Although Einstein never expressed it in these words, the cosmological constant is, in effect, an antigravity force, or antigravity field, filling the entire model universe uniformly. The number that appears in the equations could in principle have any value, provided it is constant, and Einstein labeled it with the Greek letter Lambda (Λ); but only one particular value of Λ would do the balancing act required to hold his model universe static.

But within ten years of Einstein's adding the constant to the equations, the American astronomer Edwin Hubble had established that there were other galaxies beyond the Milky Way. By the early 1930s, working with Milton Humason, Hubble had discovered the expansion of the Universe, revealed by the redshift in the light from those external galaxies. It was clear that the Universe was not static, after all, and Einstein promptly abandoned the cosmological constant, although down the years other cosmologists, more interested in the mathematics than in whether the equations describe the Universe we live in, continued to look at variations on the theme.

Like all fields, the Λ field has energy associated with it, and energy is equivalent to mass, and can distort space-time. So the Λ field can contribute to the flattening of space-time, as well as acting like anti-gravity and making the model universe they describe expand faster. These ideas were waiting in the wings when the galaxy formation experts discovered, in the mid-1990s, that in order to make their simulations match the real Universe, they needed models made up of 4 percent baryons, 26 percent clumpy cold dark matter, and 70 percent smooth something else. If the something else were Λ field, the simulations could be made to match the observed appearance of the Universe perfectly, all within the framework of the general theory of relativity. The models were dubbed ΛCDM (pronounced "Lambda CDM") and considered a great success, at least among the cognoscenti. There were a few doubts about picking up on an idea that Einstein had abandoned, and outside the modeling fraternity some astronomers weren't too sure of how accurate the modeling process was. And there was also a bizarre, but interesting, prediction from all this — so bizarre it was seldom discussed. If there really were a Λ field filling our Universe, with an overall energy equivalent to roughly twice the combined mass-energy of baryons and CDM, then its repulsive nature ought to have a noticeable effect at the fringes of the observable Universe — it ought to make the Universe expand faster as it gets older, because the antigravity effect would begin to dominate over the effect of gravity.

The point is that the cosmological constant is just that — it is *constant*. It is also very small. But gravity obeys an inverse square law, getting weaker at greater distances. When the Universe (let's not be coy about referring to models, here; this is what we think the real Universe is like) was young and matter was packed more tightly together than it is today, the effect of gravity was so strong that it overwhelmed the Λ force. But as the Universe expanded and became less dense, the effect of gravity got steadily weaker, until it became even weaker than the Λ force. From that time onward, instead of gravity slowing down the expansion of the Universe, the Λ force would operate to speed up the expansion of the Universe. But hardly anyone thought much about this implication of the ΛCDM models at first, and it was certainly not in the minds of the different teams of astronomers, from quite different backgrounds to the cosmologists and galaxy modeling fraternity, who were attempting to measure the distances to extremely remote supernova explosions at the end of the 1990s.

Attempts to extend the cosmological distance scale by measuring distances to increasingly remote objects across the Universe have an honorable tradition going back to Hubble himself. It was Hubble who discovered that the redshift in the light from a galaxy is proportional to its distance from us—but in order to discover this he had to measure distances to relatively nearby galaxies using a variety of other techniques. This calibration of the cosmic distance scale was very difficult because the more-distant galaxies appear fainter and are hard to study. The project Hubble began was eventually completed only in the late 1990s, with observations of galaxies far more remote than any he could see made with the space telescope named in his honor. There is also another subtlety. In its simplest form, "Hubble's law," as it is known, applies only to galaxies that are receding from us at less than about a third of the speed of light, corresponding to a redshift of 0.3.

For small redshifts, the shift may be thought of as the speed of recession of the galaxy divided by the speed of light—so a redshift of 0.1 means that a galaxy is receding at one-tenth of the speed of light. But a redshift of 1 does not mean that the galaxy is receding at the speed of light, because the redshift relation is actually nonlinear. As we mentioned in passing earlier, Hubble didn't notice this because his observations only extended out to redshifts corresponding to recession velocities of a few percent of the speed of light. The exact redshift-distance relationship can be calculated using the general theory of relativity (this is the prediction of the general theory that Einstein ignored when he introduced the cosmological constant), which takes account of this nonlinearity. Strictly speaking, the nonlinearity applies for all redshifts, but the correction required is too small to bother with at small redshifts. So a redshift of 2 corresponds to a recession velocity of 80 percent of the speed of light (not twice the speed of light), while a redshift of 4 corresponds to a velocity of "only" 92 percent of the speed of light. You would need a redshift of infinity to correspond to a recession velocity of exactly the speed of light. The microwave background radiation, as we have mentioned, has a redshift of about a thousand, which means that the linear size of the Universe is a thousand times bigger today than it was when that radiation was emitted, a few hundred thousand years after the Big Bang.

In the never-ending quest to measure redshifts of remote objects whose distance can be determined by other means, in the 1990s teams of astronomers began using the latest telescope technology to study the light from the stellar explosions known as supernovas. Supernovas are the biggest outbursts ever seen from ordinary stars. They happen at the end of the life of some stars when the star collapses and releases a huge burst of gravitational energy, which is converted into light and other radiation and blows the star apart. For a brief time, a single star exploding in this way radiates as much light as a whole galaxy of stars like the Milky Way (literally, brighter than a hundred billion Suns), so these beacons can be seen far away across the Universe. There are several different types of supernova, but studies of these stellar explosions in nearby galaxies whose distances are well known show that one kind, known as SN 1A (from Supernova 1A), always has the same peak brightness. This means that if such a supernova is seen in a very distant galaxy, its apparent brightness can be compared with its known intrinsic brightness to work out how far away it is. Then that direct distance measurement can be compared with the redshift to calibrate the distance scale at very great distances —in this case, "very great" corresponds to redshifts of about 1 (when the Universe was half its present size), although the record measurement of a galaxy redshift is an amazing 10.

When these observations were carried out, the researchers found that the SN 1A at very high redshifts are a little fainter than they "ought" to be, from the distances for those redshifts calculated in accordance with the general theory. One explanation for this would be that the galaxies in which those supernovas were seen are all a little farther away from us than the distances calculated from the simplest version of the general theory. But just one simple modification of the general theory was required to make the observations fit the calculations. Adding a small cosmological constant would make the Universe expand a little bit faster, carrying these remote galaxies a little farther away from us in the time since the Big Bang.

The discovery that the expansion of the Universe is accelerating made headline news in 1998 (*Science* magazine hailed it as the "breakthrough of the year") and was widely reported (even in scientific journals) in tones which suggested that cosmology had been turned on its head, and cosmologists were baffled. This came as news to many cosmologists, who were already trying to find an explanation for the "missing" 70 percent of the Universe, and to whom bringing back the cosmological constant seemed just about the simplest resolution to the puzzle. After all, the idea of the constant had been around since the time of Einstein, and was discussed in all reputable text-books of cosmology. The point, it is worth reiterating, is that like any field in physics the Λ field contains energy; energy is equivalent to mass; and a cosmological constant of the right size was just what was needed to make the Universe flat. The great thing was that a cosmological constant of exactly the same size turned out to be just what was needed to explain the high-redshift supernova observations as well. The real surprise (at least, to me) was that so many astronomers were so ignorant of the history of their own subject (even the recent history) that it took quite a while (well, about five years, which isn't that long compared with the age of the Universe) for everyone to appreciate just how well it all fits together, even if it does require a rethink about the nature and ultimate fate of the Universe.

Of course, I'm being a little harsh. Astronomers were right to be skeptical about the high-redshift supernova results until other possible explanations had been tested and independent supporting evidence had come in, even though two independent studies had found the same effect. For example, the distant supernovas might seem faint because they are obscured by dust, or the supernovas may have exploded less brightly when the Universe was younger (remember that a high redshift corresponds to a large look-back time). These possibilities have now been ruled out, largely as a result of work stimulated by the supernova studies.

Evidence that 70 percent of the mass in the Universe really is in the form of what has become known as "dark energy," and that the expansion of the Universe is accelerating as a result, has now also come from studies of the way galaxies move at the large scale, from the satellite studies of the microwave background radiation that we have already discussed, and from a beautiful technique known as the integrated Sachs-Wolfe effect which looks at the microwave radiation "shining" through the dents in space caused by the presence of clusters of galaxies, like light shining through a glass lens. A slight shift in the wavelength of this radiation, compared with the background radiation coming to us without passing through these dimples, reveals that the dents are shallower than they would be without a small antigravity effect at work even within the galaxy clusters. There is also just one observation of a supernova 1A at a redshift of 1.76, which matches the predictions of accelerated expansion but does not match other explanations, such as dimming by dust. This single observation (the first of many, it is to be hoped, at such redshifts) is exactly consistent with the idea that the expansion of the Universe slowed down for the first four or five billion years after the Big Bang and then began to speed up.* The evidence that most of what makes the Universe flat is actually dark energy is overwhelming. Which begs the question, what is dark energy?

The simplest and most natural guess is that it is indeed Einstein's cosmological constant—but this is, as yet, only an educated guess. The most important feature of the Λ field, if it does exist, is that it really is constant and has had the same strength ever since the Big Bang. In other words, because the cosmological constant is a property of space itself, the amount of this kind of dark energy in every cubic centimeter of space stays the same as the Universe expands, whereas the density of matter (light *and* dark) goes down as the Universe expands. In the fireball of the Big Bang, when the matter density was equivalent to the density of an atomic nucleus today, the cosmological constant had a negligible effect on the universal expansion. For bil-

---

*Incidentally, if the expansion of the Universe is accelerating, then the Universe must be slightly (but only slightly) older than 13.8 billion years. That age was calculated assuming no acceleration, but if the Universe is expanding faster today than in the past, it used to be expanding more slowly and must have taken slightly longer to reach its present size.

lions of years the dominant effect was that of the gravity of matter acting to slow the expansion of the Universe down. But this effect got weaker as time passed, while the cosmic repulsion associated with the Λ field stayed the same. Today, matter has thinned out to the point where it has only about half the density of the Λ field, in round terms, and the Λ field has just (a few billion years ago) started to dominate the expansion, overwhelming the influence of matter and making the expansion accelerate. It is an interesting, and possibly significant, fact (to which we shall return) that we should happen to be around to notice things like this at the only time in the entire life of the Universe when matter and dark energy are roughly in balance.

In round numbers, the density needed to make the Universe flat today is 10^{-29} grams per cubic centimeter, averaged over the entire Universe. This is equivalent to about five atoms of hydrogen in every cubic meter of space, if the atoms were all spread out evenly; but matter (both light and dark) clumps together in lumps with much greater density than this, leaving voids where there is much less density. The Λ field, by contrast, *is* spread through the Universe evenly, so there is the energy equivalent of just under 10^{-29} grams in every cubic centimeter of everything, including "empty" space. This is so small that it makes it virtually impossible to detect dark energy in the laboratory, and completely useless as a potential source of energy for human civilization — the amount of dark energy contained in the entire volume of the Earth would be just about enough to provide the annual electricity consumption of an average U.S. citizen in the year 2005. All the dark energy in a sphere as big across as the entire Solar System adds up to only the same as the energy put out by the Sun itself in three hours. But because this dark energy fills *every* cubic centimeter of the Universe, it now dominates the behavior of the Universe on large scales. It would also dominate on small scales, if there were no matter around to exert a gravitational pull. A region of space completely devoid of all matter would still contain dark energy and would stretch at an accelerating rate. If you put two tiny particles into such an empty volume, they would move apart from each other faster and faster, pushed by the "expanding spring" of dark energy.

There is one alternative to the cosmological constant that is taken seriously by some astronomers and by rather more particle physicists. This is the possibility of some form of dark energy that is *not* constant. All such candidates for the dark energy are referred to as "quintessence," because the field involved would be the fifth force field found in physics, alongside gravity, electromagnetism, and the strong and weak nuclear forces.* The important feature of quintessence is that it would have always had much the same density as matter, decreasing in step with matter density as the Universe expanded, so that it was just as important (relative to matter) in the Big Bang as it is today, and it is therefore less of a coincidence that we should be alive at a time when matter and dark energy have roughly the same density. But then, why should matter and dark energy have roughly the same density at all?

Particle physicists seldom have any difficulty dreaming up new fields and finding mathematical equations to describe them. The difficulty comes when they try to match their speculative ideas to reality. It's easy to write down a set of equations that describes a "new" quantum field filling the entire Universe and acting like a compressed spring pushing outward. A more cunning alternative to that field, which tries to explain why we should be around studying things just

---

*The name is borrowed from the ancient Greeks, who thought the material world was made of four "elements" (fire, earth, air, and water) while the universe was filled with a fifth "element," quintessence. Their quintessence was thought of as a perfect substance, which gives us our term *quintessential*.

at the time that the expansion is starting to accelerate, is called a "tracker" field. During the epoch when the universe is dominated by radiation, a tracker field "tracks" the behavior of radiation, with its own energy density falling at the same rate as the energy density of the radiation. But when the universe becomes dominated by matter, in the way we have described, the tracker field starts tracking the changing matter density instead. On this picture, the formation of stars and planets is triggered by the onset of matter domination, and with a suitable choice of properties the antigravity aspect of the tracker field starts to become important after the onset of matter domination. So it is no surprise that people, who live on planets, are around early in the era of accelerated expansion.

But the key words here are "with a suitable choice of properties." There is simply too much freedom with tracker fields, or with other variations on the quintessence theme, to pick and choose the bits that suit you. In the physicists' jargon, there are too many free parameters, and you can make quintessence fit any scenario. The whole thing is convoluted, contrived, and implausible. There are similar difficulties with an otherwise appealing suggestion from the M-theorists, who have contrived models in which instead of antigravity accelerating the expansion of the universe, gravity itself (in the form of gravitons) leaks away out of our "brane" more and more as time passes, so that gravity gets weaker as time passes, loosening its grip on distant galaxies. By contrast, the cosmological constant is simple, and a natural (some would say, inevitable) part of Einstein's general theory of relativity. The only free parameter in the equation is the choice of the right size for the energy density of the Λ field to match the observed acceleration of the Universe and the flatness of space-time. The key puzzle then is, Why is the energy density of the Λ field so small?

In quantum field theory there is no problem explaining why

"empty space" should contain energy. The problem is explaining why it doesn't contain a lot more. Such energy, called the energy of the vacuum, arises naturally in the context of grand unified theories and supersymmetry. There is nothing in any of this which says that the vacuum energy must be zero, only that it must be the same everywhere. We can go back to the analogy of the mountain lake. It's as if theory says that the surface of the lake must be flat, but it doesn't have to be at sea level. The problem is that the "natural" energy scales involved are vastly bigger than the energy density required to explain the flatness of the Universe. Such a vacuum energy associated with quantum gravity, for example, would produce an energy density of 10^{108} in the particle physicists' usual electron volt units. That's the kind of vacuum energy that drove inflation, through the same kind of processes that are at work in the Universe today but on a vastly less powerful scale. Some cosmologists would say that the Universe is literally experiencing a weak form of inflation today—which sounds clever, and may even be true, but doesn't actually add anything to our understanding. If anything, it only highlights the puzzle that this "inflation" is so weak today. Even at the energy level associated with the grand unified theories, the energy density would be 10^{96} in the same units, while the smallest "natural" vacuum energy, according to the established versions of supersymmetry, would be "only" 10^{44}. In these units, the actual value of the vacuum energy density is 10^{-12}. In other words, even the smallest "natural" vacuum energy density is 10^{56} times bigger than the observed vacuum energy density.

Apart from any other consideration, a vacuum energy as large as even supersymmetry suggests would produce an antigravity effect that would rip the material Universe apart before there was any hope of irregularities like galaxies, stars, and planets forming. So it's hardly surprising that until the SN 1A observations came along, most particle theorists assumed that some universal energy-suppression mechanism had forced the vacuum energy all the way down to zero, rather than leaving just a tiny residue. The existence of a dark energy density that is not zero but is very small indeed is a real puzzle for the theorists. But there is one intriguing line of attack on the problem which may yet bear fruit. The actual vacuum energy is roughly the right size to match any as yet unobserved symmetry-breaking associated with energies corresponding to masses of a few thousandths of an eV—just the masses now known to be associated with neutrinos. This is, as yet, no more than a tantalizing hint at a possible resolution of the problem, but just maybe the coincidence is pointing us toward a Deep Truth about the Universe.

Many of these questions should be resolved over the next twenty years or so, by new generations of satellites that will observe distant supernovas in their thousands, and by ground-based experiments at places like CERN and the Boulby mine. The "New Standard Cosmology," as it is sometimes called, can be summed up in five statements:

- The Universe we live in emerged from an early epoch of rapid expansion (inflation), then slowed its expansion rate.
- The Universe today is flat and its acceleration is expanding.
- The irregularities in the Universe today (galaxies, stars, and all the rest, including ourselves) result from quantum fluctuations during inflation.
- The Universe is made up of roughly 70 percent dark energy and 30 percent matter.
- The matter in the Universe is made up of roughly seven times more nonbaryonic dark matter than baryonic matter, with only 10 percent of the baryonic matter (0.4 percent of the total mass-energy of the Universe) in the form of bright

stars. Neutrinos contribute as much mass, overall, as bright stars.

So we can answer the question posed in the title of this chapter. It is mostly dark energy that holds the Universe together today; but, seemingly paradoxically, if the acceleration of the expansion continues, it is also dark energy that will ultimately blow the Universe apart.

But where do we fit into all this? Is it just a coincidence that intelligent life has emerged in the Universe early in the epoch of accelerated expansion, or is this telling us a Deep Truth about the nature of the Universe?

We certainly do live at a special time in the life of the Universe. Taking the simplest cosmological constant explanation of the accelerated expansion (as Ockham's razor would encourage us to do), ten billion years ago, at a redshift of 2, the dark energy contributed only 10 percent of the density of the Universe, but in ten billion years' time, the dark energy will contribute 96 percent of the total density. At earlier and later times the difference is even bigger — at the time of recombination, for example, the matter density was a billion times greater than the dark energy density. The fact that the dark energy today contributes roughly the same as matter (within a factor of 2) to the density of the Universe really is odd. But at least this helps the observers. The switch from a decelerating Universe dominated by matter to an accelerating Universe dominated by dark energy happened between redshift 0.1 and redshift 1.7, conveniently close to us to make detailed observations of the changeover feasible with the next generation of satellite detectors.

Meanwhile, theorists continue to puzzle over the similar sizes of the contributions of dark energy and matter to the flatness of the Universe. If the cosmological constant really is constant, this is equivalent to asking why the density of dark energy is so small. Because the energy density is so small, the Λ field had very little effect early in the expansion of the Universe, which allowed stars, galaxies, and clusters of galaxies to form by gravitational collapse even though the Universe was expanding. It took time, as we shall see in the next chapters, for the first stars to run through their life cycles and seed the galaxies with the raw materials necessary for the formation of planets and life. It then took more time for intelligence to emerge on at least one of those planets. By the time all that had happened, the matter density of the Universe had dropped below the dark energy density, and the acceleration of the expansion was just beginning to become noticeable. In the not-too-distant cosmological future, though, the runaway expansion of the Universe may make life impossible, and in any case there will be nothing left to see. People are interesting and complicated entities, and we live in the most interesting and complicated time in the life of the Universe because that is the only time when creatures like ourselves could exist.

But all of this applies only for very small dark energy densities. For any other values, interesting and complicated things like ourselves could never exist. If the dark energy density were large, it would have overwhelmed the gravitational effect of matter in the early Universe, and spread matter increasingly thinly, in a runaway expanding universe in which stars, planets, and people would never have formed.

The other extreme possibility is that the Λ term could be *negative*. If a positive Λ term corresponds to dark energy with antigravity, a negative Λ term corresponds to dark energy with an extra kind of positive gravitational force. With such an effect adding to the attraction of matter, such a universe would have collapsed in upon itself too quickly for stars, planets, and people to form. So the puzzle of the size of the cosmological constant is similar to the puzzle that baffled a previous generation of astronomers, of why the Universe should be flat, exactly balanced on the knife-edge between runaway expansion and precipitate collapse. The resolution of that puzzle turned out to be a completely new idea, inflation. I suspect that a resolution of the puzzle of the cosmological constant will also turn out to be something completely new, which nobody has yet imagined and which will tell us a new Deep Truth about the nature of the Universe. Until that great new idea emerges, though, the best explanation of the "coincidence" comes from an idea known as anthropic cosmology. Some scientists regard this as a council of despair. I rather like it; and it is certainly the best explanation we have as yet of why the Universe seems such a comfortable home for life-forms like ourselves.

The basic idea behind the anthropic principle is that there is much more — perhaps infinitely more — to the Universe than we can see. Not more dark stuff in the gaps between the bright stars, but more space-time beyond the limits of the observable Universe. This "super-universe" can be referred to as the Cosmos, to avoid confusion. If space-time (the Cosmos) is infinite, then our expanding Universe could be just one bubble within that infinite sea, and there could be many (infinitely many) other bubble "universes" out there, each formed in its own version of inflation, forever unseeable and un-touchable from our Universe. Just as we have learned that our Solar System is not unique, and our Milky Way Galaxy is not unique, per-haps it is time to appreciate that our Universe may not be unique. Some versions of inflation actually imply that there must be infinitely many bubble universes of this kind in an infinite sea of space-time, liker separate bubbles forming in a bottle of fizzy drink (an infinitely large bottle!) when the top is unscrewed. The anthropic argument says that there is nothing in the laws of physics which specifies what value the cosmological constant ought to have — exactly equivalent arguments apply to other "constants" of nature, but this is not the place to go in to the anthropic cosmology story in detail—so the constant will have different values in different bubbles.

In some bubbles (some universes) the constant will be large, expansion will accelerate from the beginning, and there will be no stars, planets, or people. In other bubble universes, the constant will be negative, and the bubbles will collapse before interesting things like life can happen within them. Only in bubbles which have a small cosmological constant and are otherwise "just right" will life emerge. Overall, there is a vast variety of possible universes, and we live in an anthropically permitted location within this Cosmos.

Although the ideas of anthropic cosmology have received a boost from the theory of inflation and the discovery of an otherwise surprisingly small cosmological constant, they actually have a long history. In its modern incarnation, serious anthropic cosmology jumps off from the work of the British researcher Brandon Carter in the early 1970s, although the great astrophysicist Fred Hoyle had used one specific anthropic argument to make a key discovery (which we discuss in the next chapter) in the 1950s. Carter pointed out at a meeting in Poland in 1973 that "what we can expect to observe must be restricted by the condition necessary for our presence as observers," which is still the best succinct statement of the anthropic principle. But even as long ago as 1903, in his book *Man's Place in the Universe,* Alfred Russel Wallace (best known as the independent discoverer of "Darwin's" theory of evolution by natural selection) wrote: "A vast and complex universe as that which we know exists around us, may have been absolutely required . . . in order to produce a world that should be so precisely adapted . . . for the . . . development of life."

But you don't need to look at the entire Universe to get a feel for anthropic reasoning. In our Solar System, for example, it probably is pure chance, with no anthropic significance, that there should be four rocky planets (Mercury, Venus, Earth, and Mars) orbiting close to the Sun, rather than three, or five. There is no fundamental reason why the evolution of astronomers on Earth could only have happened in the presence of three other rocky planets. But there is a fundamental reason why the evolution of astronomers happened on Earth rather than on the other three rocky planets. A kind of anthropic "selection effect" has been at work in our own Solar System. Of the three neighboring planets — Venus, Earth, and Mars — only Earth offers a suitable home for life-forms like ourselves today. Life-forms like us can survive only on Earth, so when we look around us it is no surprise to find that Earth is the planet we live on. *Provided* that there is in some sense a choice of universes, with different physical properties, exactly the same logic says that life-forms like ourselves will only be around to notice what is going on, and measure things like the cosmological constant, in universes that are suitable homes for life-forms like ourselves — and this is more than mere tautology, as the planetary example shows. Since we are alive, it is no surprise, the argument runs, to find that we live in a Universe conducive to life, any more than it is a surprise to find that fish live in water.

It is largely a matter of personal preference whether you like this idea or abhor it, and even those who like it would be happier if it turned out that there is some fundamental reason why the dark energy density is as small as it is. The bottom line, though, is that we *do* exist, and we live in a flat Universe about 14 billion years old in which the dark energy has recently started to overwhelm gravity and increase the expansion rate. Given the nature of the Universe, how did we come to be here? To answer this question, we first need to know where the stuff we are made of, all the variety of baryonic material apart from the hydrogen and helium that emerged from the Big Bang, came from. We can now ignore the Universe at large and concentrate on what went on in our own Galaxy (and the hundreds of billions like it) when the Universe was young.

7

Where Did the Chemical Elements Come From?

Although Alfred Russel Wallace, writing more than a hundred years ago, had no idea of the true extent and complexity of the Universe, his remarks about the relationship between life on Earth and the Universe at large resonate today. In another example of anthropic reasoning, it seems entirely possible that a very large universe billions of years old is a requirement as the "stage" on which life-forms like ourselves can perform. The fact that we exist means that when we look out into the night sky we must inevitably see a large, old Universe.

The reasoning runs like this, starting from the facts that the Universe is flat, expanding, has a small cosmological constant, and contains irregularities caused by matter clumping together under the influence of gravity. How did some of those clumps of matter form into stars, planets, and people? This order is important because, as I have emphasized elsewhere (in *Stardust*), life begins with the process of star formation. We are made of a variety of baryonic material, not just hydrogen and helium — in fact, not helium at all. Every element in your body apart from the atoms of hydrogen has been manufactured inside a star, and that takes time, time during which the Universe kept on expanding. So the fact that we exist requires that the Universe be large and old.

The modern understanding of the way in which the chemical elements were manufactured inside stars is another archetypal example of the power of combining what we know of physics on the large scale — in this case, the scale of stars — with what we know of physics on the small scale — in this case, the scale of atomic nuclei. This time, the study of the physics of stars — astrophysics — pointed the way to one of the key features of quantum physics, the uncertainty associated with wave-particle duality.

To a physicist, a star viewed from outside is a simple thing. It is a ball of stuff held together by gravity and prevented from collapsing further by the heat generated in its core, which sets up an outward pressure that balances gravity. If you know how bright a star is and how massive it is, it is a trivial calculation (really — secondary school stuff) to work out how hot it must be in its heart to prevent collapse. It doesn't matter what the star is made of, or where it draws energy from, it has to have a certain internal temperature to provide the pressure to resist the pull of gravity and to shine as brightly as it does. The Sun is a fairly ordinary star, and close enough to be studied in some detail, so it was the first star investigated in this way; but thanks to spectroscopy astronomers can measure the temperatures of other stars, and thanks to the way stars in binary pairs orbit around each other, in many cases astronomers can also measure their masses.

In the 1920s, astrophysicists were able to carry out the simple calculation and determine that the central temperature of a star like the Sun must be about 15 million K. The only plausible source of the energy required to keep the Sun shining was the conversion of mass into energy in line with Einstein's equation $E = mc^2$ — but where could the mass m come from? By then, the technology of particle physics was already sophisticated enough to measure the masses of atomic nuclei with some accuracy, and it was clear that mass could be (must be) "lost" if light nuclei fused together to make heavier nuclei. For example, a nucleus of helium-4, consisting of two protons and two neutrons, has a mass of 4.0026 units (on a scale where the mass of a carbon-12 nucleus is defined as 12 atomic mass units, or amu), but the total mass of four individual protons is 4.0313 amu. If four protons (hydrogen nuclei) could be persuaded to combine to form one helium nucleus, 0.0287 amu of mass would be liberated in the form of energy — just over 0.7 percent of the total mass of the four original protons.*

But there's a snag. If four protons could somehow be squeezed closely enough together, the strong force would dominate and bind them tightly together, with two electrons being ejected (the process known as beta decay), to make a single helium-4 nucleus. It is, indeed, the strong force that holds such nuclei together even though the positive charge on all the protons repels them from one another. But the strong force has a very short range. When two protons approach each other, the mutual repulsion they feel as a result of their positive charge becomes strong enough to scatter them away from each other long before the strong force gets a chance to act, except under very extreme conditions. In order for two protons to get close enough together for the strong force to take over and make them stick, ejecting a single positron to form a deuteron, the nucleus of deuterium,

---

*If we were being scrupulously accurate, we would also allow for the masses of the two electrons produced when two of the protons are converted into neutrons; but the mass of each electron is only 0.05 percent of the mass of a proton, so the argument still stands.

they have to be moving very fast indeed, which means they have to be in an environment of high temperature and pressure. Such conditions existed, as we have seen, in the first few minutes of the Big Bang, although that wasn't known in the 1920s. What *was* known in the mid-1920s was that at a temperature of "only" 15 million degrees nuclei would not be able to fuse in the way required to provide the energy source of the Sun and other stars, according to the laws of physics known at the time. It was the discovery of quantum uncertainty — new physics — that resolved the dilemma.

Quantum uncertainty tells us that an entity such as a proton does not have a precise location at a point in space but is spread out in a fuzzy fashion. In terms of wave-particle duality, you can think of this as an aspect of the wave nature of a "particle." Waves are intrinsically spread-out things. So when two protons approach each other, it is possible for their waves to begin to overlap even though the old physics says that they are not yet touching. As soon as the waves mingle in this way, the strong force can get to work, tugging the protons into a tighter embrace and (with the aid of the weak interaction) forcing the ejection of an electron. This process is sometimes called the "tunnel effect," because the electrical repulsion between two positively charged particles is an insurmountable barrier according to classical physics, and the protons seem to tunnel through the barrier with the aid of quantum uncertainty (it also works for other particles, of course). When quantum uncertainty came on the scene in the late 1920s, it turned out that it provided just enough opportunity for protons to get together in the heart of the Sun to liberate the energy required to keep the Sun shining.

It is, perhaps, worth spelling out just how extreme the conditions under which these nuclear interactions take place are, compared with everyday life on the surface of the Earth. The core of the Sun, the region in which nuclear reactions take place and convert matter into energy, extends only a quarter of the way from the center to the surface, which means that it occupies only 1.5 percent of the volume of the star. Within that volume, it is far too hot for electrons to be held in an electromagnetic embrace with nuclei to form atoms, and the nuclei are packed in with a density 12 times that of solid lead on Earth, or 160 times the density of water. The pressure in the core is 300 *billion* times the atmospheric pressure at the surface of the Earth. Because of its high density, the inner 1.5 percent of the volume of the Sun actually contains half its mass. The temperature, as we have mentioned, reaches a peak of some 15 million K (the temperature at the outer edge of the core is about 13 million K), yet nuclei are so much smaller than atoms that even at these densities they behave almost exactly in the way atoms of a gas behave under less extreme conditions, moving rapidly and bouncing off one another in repeated collisions. The core of the Sun is as good an approximation as physicists are likely to find of their concept of a "perfect" gas.

These conditions are extreme by human standards, but much less extreme than the conditions that existed in the Big Bang. The crucial difference is that the conditions at the heart of a star stay the same for millions, even billions, of years; the Big Bang was over in a few minutes, and there simply was not time for nucleosynthesis to proceed very far. Even under the conditions that exist at the heart of the Sun, it is only in extremely rare head-on collisions that two protons will get close enough to each other for tunneling to work and for the strong force plus a variation on the beta decay process to create a deuteron, which consists of a single proton and a single neutron bound together.

It took more than twenty years for all the details of the nuclear fusion processes that keep the Sun shining to be worked out. For example, the details of the way protons interact with one another when they collide (their "cross-sections") were measured in particle accelerators and these measurements were used to calculate how often such collisions result in the production of deuterons in the heart of the Sun. But we can skip to the results of all that effort. Although the protons are bouncing around colliding with one another many, many times every second, on average an individual proton will only meet and fuse with a partner after a billion years. So for every two billion protons you start with, after one year just a single a pair of them will have gotten together to form a deuteron.

Once this happens, within about a second a third proton sticks to the deuteron to make a nucleus of helium-3. When other protons collide with the helium-3 nucleus, they just bounce off, and there are fewer helium-3 nuclei around than there are protons so they collide with each other less frequently. But when they do collide, they are relatively eager to fuse; after only about a million years of wandering through the core of the Sun, the helium-3 nucleus will meet another helium-3 nucleus and combine with it to form a nucleus of helium-4, with two protons being ejected.

This specific chain of events is called the proton-proton chain, and the overall process of converting hydrogen into helium (by whatever means) is sometimes referred to as hydrogen burning. The net result, after a little over a billion years, is the conversion of four protons into one helium-4 nucleus with the release of energy.* For each helium-4 nucleus manufactured in this way, 0.048×10^{-27} kilograms of mass is "lost." But there are so many billions of particles in the core of the Sun, and so many fusion reactions occurring every second, that over-all the Sun "loses" 4.3 million tonnes of mass every second, by con-

---

*There are some "side chains" to this process, involving slightly different interactions, as when a helium-3 nucleus meets a helium-4 nucleus, but these only contribute a small amount of the energy released.

verting 600 million tonnes of hydrogen into just under 596 million tonnes of helium. (The amount of hydrogen transformed into helium inside the Sun each second is roughly the same as the amount of hydrogen in the water of North America's Lake Michigan.) It has been doing this for roughly 4.5 billion years, and in the process it has so far released as energy only a few hundredths of 1 percent of its original mass of hydrogen. It should be no surprise that the percentage figure is so low, since each helium-4 nucleus produced is associated with the release of only 0.7 percent of the mass of the four original protons. Even if the Sun were entirely made of hydrogen (which it is not), and all that hydrogen were converted into helium-4, the mass it would lose in the process would only be 0.7 percent of its original mass. What those figures really tell you is just how big the Sun is!

The speed with which all this occurs is self-regulating. If the Sun shrank a little bit, the pressure and temperature at its heart would go up, and more fusion would occur, releasing more energy. The increased heat would make the star swell, lowering the pressure and cooling things down again. If the star expanded, it would cool in its heart, the energy production rate would slow down, and it would shrink back to its stable size. But when the hydrogen fuel available in the core runs out (as it will in the Sun in about four billion years), everything has to adjust to a new stable balance.

Just as hydrogen nuclei (protons) can be fused to make helium-4 nuclei with energy released along the way, so helium nuclei can be fused to make nuclei of other elements, although proportionately less energy is released in the process. But these processes of stellar nucleosynthesis operate at higher temperatures than the proton-proton chain, and they cannot occur as long as hydrogen burning is going on. Although it may seem paradoxical, hydrogen burning actually keeps the core of a star like the Sun relatively cool. When all the hydrogen fuel is exhausted, the first thing that happens is that the pressure in the core drops, allowing the star to shrink. This releases gravitational energy, which heats the core until a point is reached where new fusion reactions can take place. The energy released by these reactions then stabilizes the star with a higher internal temperature and pressure for as long as the new source of fuel lasts.

Because the helium-4 nucleus is a particularly stable configuration of nucleons, it acts like a single unit in many of these interactions, and it is sometimes called the alpha "particle"; in the next stage of stellar nucleosynthesis heavier elements are largely built up by sticking alpha particles together. Some nuclei may then absorb more protons, and some may eject particles to make nuclei of other elements or isotopes, but by and large elements whose nuclei contain multiples of four nucleons (such as carbon-12 and oxygen-16) are particularly stable and common compared with other heavy elements. (To an astronomer, everything except hydrogen and helium is a "heavy" element; astronomers also refer to everything except hydrogen and helium as "metals," presumably just to annoy chemists.)

You might guess that the next step in the fusion process, then, would be the formation of beryllium-8 nuclei from pairs of helium-4 nuclei. But beryllium-8 is the exception that proves the rule. It is very unstable indeed, and if two helium-4 nuclei do happen to collide in the right way to stick together, they do so only for a very brief instant. This posed a serious puzzle for astrophysicists, since there are an awful lot of elements heavier than beryllium in the Universe, and nowhere to make them except inside stars. How could stellar nucleosynthesis skip over beryllium and make the heavier, more stable nuclei? The puzzle was only resolved in the 1950s, when Fred Hoyle came up with an insight which showed how stable carbon-12 could be made inside stars out of three helium-4 nuclei, in an interaction known as the triple-alpha process. Although this book concentrates on twenty-first-century ideas rather than the ancient history of the 1950s, Hoyle's insight is so profound, and so relevant to modern cosmological thinking, that it is worth a small digression to examine its importance. It was, in fact, the first (and remains the most success-ful) application of anthropic reasoning to make a prediction about the nature of the physical world.

The lifetime of a beryllium-8 nucleus is only 10^{-19} seconds, but even in that short time, under the conditions that prevail in the heart of a star where hydrogen burning has come to an end, there is time for some of the beryllium nuclei produced to collide with an alpha parti-cle. The problem is that because beryllium-8 is so unstable, such a collision ought to blow the nucleus apart, not stick the three alpha particles together.

Hoyle reasoned that since carbon exists (most notably, from the human point of view, in carbon-based organisms like ourselves) there must be something in the laws of physics which allows the third alpha particle to stick in spite of the instability of beryllium-8. In his own words at the time, "since we are surrounded by carbon in the natural world and we ourselves are carbon-based life, the stars must have discovered a highly effective way of making it, and I am going to look for it" (Mitton, *Fred Hoyle*). If the triple-alpha process did do the trick, it would make a nucleus of carbon-12—but could there be something special about the kind of carbon-12 nucleus it produces that prevented its immediate disintegration?

From quantum physics, Hoyle knew that atomic nuclei usually exist in their lowest energy state (called the ground state), but under the right conditions they can absorb a quantum of energy (such as a gamma ray photon) and enter a so-called excited state. After a short time, they then emit a gamma ray photon and settle back down into the ground state. This is similar to the way in which an atom can absorb a photon of light, with an electron jumping up to a higher energy level, then re-emit the light as the electron settles back into a lower energy state. A good analogy is with the way the string of an instrument such as a violin or a guitar can vibrate in a basic way to make its natural (fundamental) note, but when plucked in the right way it will vibrate in a harmonic, making a higher note.

Hoyle concluded that the only way in which three alpha particles could get together to make a single nucleus of carbon-12 under the conditions that prevailed inside the stars of interest would be if the carbon-12 nucleus had a natural "resonance" corresponding (to within a few percent) to the energy of a beryllium-8 nucleus in its ground state *plus* the energy of an incoming alpha particle with the appropriate temperature. Then the kinetic energy of the incoming alpha particle would all go into "exciting" the carbon-12 nucleus, and none would be left over to blow it apart. The excited carbon-12 nucleus could then radiate a gamma ray photon and settle down into its ground state in the usual way. But the trick only worked if the energy needed to make the excited state of carbon-12, the resonance, was a tiny bit less than the energy of the incoming alpha particle. If it were even a tiny bit more, the incoming alpha particle would not have enough energy to do the job; if it were a lot less, there would be enough leftover kinetic energy to blow the nucleus apart anyway.

A resonance with about the right amount of energy had been suggested by some experiments at the end of the 1940s, but nobody had made the connection with processes operating inside stars at that time. When Hoyle raised the matter with the particle physicists at Caltech a few years later he was told that the latest experiments suggested that the earlier work was in error. Hoyle refused to believe this, and pestered the experimenters to repeat the measurements once again. In effect, Hoyle was saying that the fact that we exist means that the laws of nature *must* be such that the nucleus of one specific element, carbon-12, has an excited state with a specific energy. In the 1950s, this chain of reasoning seemed preposterous to most scientists. But like all good scientific ideas, it could be tested.

The way to test it was to measure the properties of carbon-12. The technology to do this existed at the Kellogg Radiation Laboratory in California. Hoyle made the precise prediction that carbon-12 must have a resonant excited state with an energy exactly 7.65 MeV above its ground state, and with some difficulty he persuaded the researchers at the lab to carry out the experiments to search for this excited state of carbon-12. Willy Fowler, who led the team that carried out the tests, later told me that he only did so in order to get Hoyle to shut up and go away; he never expected to find the resonance. The entire experimental team was astonished when it turned out that Hoyle's prediction was accurate to within 5 percent. Carbon-12 really does have a resonance in just the right place to make the triple-alpha process work. *Why* this should be so is as much a mystery as why the Universe is expanding at just the right rate to allow for the formation of stars, planets, and people; the same anthropic solution — that there is a multitude of universes out there with all kinds of physical laws, and we only exist in the one which is just right for us — is one possible solution to the puzzle. But what matters from a practical point of view is that stellar nucleosynthesis can leap over the beryllium-8 gap, and from then on the buildup of heavier elements inside stars is pretty much plain sailing. There is no better example of the way in which astrophysics and particle physics combine to give us a deep understanding of the nature of the Universe.

Once carbon nuclei exist inside a star, it is relatively easy to make heavier elements, up to a point. Helium burning continues in the core of the star, with a slightly cooler shell of unburned helium surrounding it (and hydrogen in the outermost layers of the star) until all the fuel is exhausted. Then the core shrinks once again, making it hotter and enabling reactions to occur in which carbon-12 nuclei fuse with alpha particles to make nuclei of oxygen-16. This stabilizes the star so long as the carbon fuel lasts, when the processes of contraction, heating, and the triggering of a new wave of fusion repeat. Each successive step requires higher temperatures, of course, because the more protons there are in an atomic nucleus the more positive charge it has, and the faster an incoming alpha particle (itself positively charged) has to move to tunnel through the barrier and reach the nucleus. In this way, elements such as neon-20, magnesium-24, and silicon-28 are produced, and an old star may have a series of shells, like onion skins, surrounding its core, with heavier elements at the center and lighter elements nearer the surface. Experiments using particle accelerators confirm that under these conditions elements such as fluorine-19 and sodium-23 can be produced in modest quantities when the more common nuclei with masses divisible by four interact with particles from their surroundings, absorbing the odd proton and emitting the odd positron. The great triumph of all this work is that the understanding of such nuclear interactions developed here on Earth, combined with the understanding of stars developed by astrophysicists, predicts the same proportions of the various elements that are actually observed in the spectra of the Sun and other stars. This is beautifully demonstrated by the detailed understanding astrophysicists have developed of one especially important set of nuclear interactions.

Provided there are already traces of carbon and oxygen around, there is a second way, apart from the proton-proton chain, in which stars slightly heavier than the Sun and with slightly hotter cores can burn hydrogen to make helium. Because this process involves nitrogen and oxygen as well as carbon, it is often referred to as the CNO cycle; its special feature is that although all these nuclei are involved in the cycle, once it has reached equilibrium they are not "used up," and the net effect of each traverse of the cycle is to convert four hydrogen nuclei (protons) into one helium-4 nucleus (an alpha particle).

The cycle works like this. A carbon-12 nucleus captures a proton and becomes a nucleus of nitrogen-13. The nitrogen-13 nucleus spits out a positron and becomes carbon-13, which then picks up a proton and becomes nitrogen-14. The nitrogen-14 absorbs a proton to become oxygen-15, the oxygen-15 ejects a positron to become nitrogen-15, and finally the nitrogen-15 absorbs a proton and immediately throws out an alpha particle, leaving carbon-12, ready for the cycle to repeat.* Every step in this process has been studied in particle laboratories on Earth, so we know just how fast they proceed and under what conditions. This has led to a profound discovery of major importance to ourselves and to all life on Earth.

The slowest reaction in the cycle is the one that converts nitrogen-14 into oxygen-15. As a result, early in the life of a star big enough to trigger this kind of hydrogen burning far more carbon is being converted into nitrogen than the amount of nitrogen being converted into oxygen. (But remember, carbon is only a minor ingredient in such a star, which is still mostly made of hydrogen and helium at this stage of its life.) As more nitrogen builds up, even though each nucleus takes its time before interacting there are a lot of nuclei to interact, so eventually an equilibrium is established. It's a bit like the way you can get the same amount of water out of a single tap opened wide as from the many

---

*There are other, less important side chains that feed into the cycle, which gives astronomers the chance to indulge their taste for puns by referring to the CNO "bi-cycle" or even the CNO "tri-cycle," but these are not important for our discussion here.

holes in a garden sprinkler attached to the tap by a hose—if you just had one tiny hole in the sprinkler, it wouldn't be anywhere near so effective. This means that although for a single circuit round the cycle the net effect is to convert four hydrogen nuclei into one helium nucleus, with energy being liberated along the way, over the lifetime of the star a side effect is to convert carbon into nitrogen.

Why is this so important? Because nitrogen is one of the essential elements of life (that is, life as we know it here on Earth), and the CNO cycle is the *only* mechanism in the Universe for manufacturing nitrogen. There are other ways to make carbon and oxygen, as we have seen, but not to make nitrogen. We can say with absolute certainty that every bit of nitrogen in your body was made inside stars like the Sun, but probably a bit more massive than our Sun, by the CNO cycle. Without the CNO cycle, we would not be here; it really is true that life begins with the process of star formation.

Indeed, we would not be here if these elements manufactured inside stars did not escape into the Universe at large and form the raw material for later generations of stars, planets, and people. We are coming to that part of the story. But first, we left the story of stellar nucleosynthesis itself at silicon-28, and it is time to pick up the threads and discuss how even heavier elements are made.

We left the story at silicon-28 for a good reason—the simple step-by-step addition of alpha particles to heavier nuclei, steadily increasing their mass four units at a time, stops there. Things become complicated because the core of the star is now so hot (about 3 billion K) and dense (millions of grams of material packed into every cubic centimeter) that dense nuclei sometimes get smashed apart. A single silicon-28 nucleus, for example, might "photodisintegrate" to release seven helium-4 nuclei. But this flood of extra alpha particles then combines with other silicon-28 nuclei, perhaps with more than one alpha particle being absorbed by a single silicon-28 nucleus, making sulfur-32, chlorine-36, argon-40, and heavier nuclei in a single step. Occasionally, all seven of the released alpha particles may be captured by a single neighboring nucleus of silicon-28, turning it into nickel-56 in one step.

But this is nearly the end of the line. Nickel-56 is unstable, and quickly spits out a positron, converting itself in to cobalt-56; the cobalt-56 nucleus itself ejects another positron, settling down into a stable state as iron-56. Indeed, iron-56 is the most stable nucleus there is, in which the nucleons (twenty-six protons and thirty neutrons) are bound together more tightly than in any other nucleus. This means that it is the end of the road, as far as fusing lighter nuclei together to make heavier elements and releasing energy in the process is concerned. The only way to make elements heavier than iron is to put energy *in* — the nuclei have to be forced together by some outside process, drawing on a large source of energy. The only energy source that can do the job is gravity, and even gravity is not strong enough for the task unless a star has much more mass than our Sun.

Very many stars never even get to the point of manufacturing elements as heavy as silicon, sulfur, chlorine, and the rest up to iron. The Sun is a very ordinary star, still burning hydrogen to helium, which will eventually get hot enough in its core to burn helium into carbon, with perhaps a little nitrogen and oxygen being manufactured in the CNO cycle along the way. But when helium burning ends, a star like the Sun cannot contract enough to get hot enough inside to start burning carbon into oxygen. It will shrink in upon itself and cool down, eventually becoming a solid ball of carbon (or if you are romantically inclined, a single diamond crystal) surrounded by a layer of helium and a trace of hydrogen. It will have become a white dwarf, no bigger than the Earth but still containing a large fraction of its original mass. Only 10 percent of stars are more massive than the Sun, but they are crucial in accounting for the origin of the elements. A star with more than about four times the mass of our Sun is needed for carbon burning to occur, and in order to make all the heavy elements you need to start out with a star of at least eight or ten solar masses. Crucially, though, all these stars, even one as small as the Sun, do not hold on to all the material they start out with.

At the various times in its life when the core of a star contracts and gets hotter — for example, at the start of helium burning in a star like our Sun — the extra heat from the core makes the outer layers of the star expand. This results in at least a quarter of the mass of the star (if it starts out with the same mass as our Sun) being blown away entirely into space, forming an expanding cloud of material escaping into the Galaxy at large. Such clouds are among the most beautiful objects in the Universe. They are known as planetary nebulae, because in telescopes of a bygone era, with relatively little light-gathering power, they looked a little bit like planets. But modern instruments show them in a variety of colorful shapes, reminiscent of flowers, butterflies, and glowing rings of light, among others. The complicated series of interactions that produces iron-56 (and it is even more complicated than the simple outline we have sketched here) simply blows the outer layers of a star away, if it ever gets to that stage of nucleosynthesis. Although the Sun will lose no more than a third of its mass over its lifetime, a star which started out with about six times the mass of our Sun may eject five solar masses of processed material, laced with heavy elements, into space before the core settles down as a white dwarf with about the same mass our Sun has today.

If, however, a star starts out with a little more mass still — six to eight times the mass of the Sun — the calculations show that it must be completely disrupted at the end of its life. This is because the remnant left over at the end of the star's active life has too much mass to become a stable white dwarf. The critical mass is about 1.4 solar masses, and it is known as the Chandrasekhar limit, after the astronomer who first calculated it, Subrahmanyan Chandrasekhar. A star that is no longer supported by the outward pressure resulting from nuclear fusion and has more mass than the Chandrasekhar limit must collapse under its own weight, and cannot form a stable white dwarf. As it collapses, the core of such a star gets hotter — remember that nuclear fusion keeps a star *cool* by preventing collapse — so hot that carbon can "burn" in a variety of interactions to make heavier elements, releasing energy in the process; but the gravitational pull of the star is too small to hold on to the fragments from the resulting explosion.

Apart from the simple addition of a helium-4 nucleus to a carbon-12 nucleus to make oxygen-16, when the density, pressure, and temperature are high enough, carbon nuclei can interact directly with one another in a variety of ways; the simplest is when two carbon-12 nuclei fuse and eject an alpha particle (which then goes on to interact with another nucleus) leaving behind a nucleus of neon-20. Each such event actually releases more energy than the fusion of three alpha particles to make a single carbon-12 nucleus; this explosive carbon burning stimulates the fusion of nuclei right up to iron-56, with all of this material scattered into space as the star explodes. At least half a solar mass of iron and about an eighth of a solar mass of oxygen will be among the elements spread across the Galaxy in such an explosion, the simplest kind of supernova. But it still doesn't make any elements heavier than iron. To do that, you need to start out with an even more massive star.

Stars which start out with masses more than eight or ten times that of the Sun end their lives in an even more spectacular fashion, and are the source of all the elements heavier than iron, including gold, uranium, lead, mercury, titanium, strontium, and zirconium. There is no need to go into all the details here — not least since I covered that ground in *Stardust* — but what matters is that even after losing mass from its surface at earlier stages in its life, such a star will still have a sizable amount of material in its outer layers, above the core where this kind of collapse and explosion take place. And the core itself will be large enough that it will still have a strong enough gravitational pull to hold itself together when it does collapse and release gravitational energy. When nuclear burning in the core of such a star ends and can no longer support the weight of the star, the core will have a mass greater than the Chandrasekhar limit, and will collapse, triggering an explosive release of energy without being entirely disrupted.

For the outer layers of the star, many solar masses of material, it will be as if the floor has been taken away from beneath them, leaving an almost bottomless pit. The outer layers of the star — all those solar masses of material — will start to fall inward, only to meet the blast wave from the core explosion moving outward. The blast will squeeze and heat the material of the outer part of the star in a shock wave, where conditions are so extreme that neutrons (released in interactions which disrupt some nuclei) will be forced to fuse with heavy nuclei, building up the heavier elements than iron. In fact, some of these heavier elements will already have been synthesized in the extreme conditions of the collapsing core itself, ultimately drawing on the gravitational energy released by the collapse to provide the squeeze which forces the nucleons together, and the job is finished off in the shock wave.

The shock is also boosted by a flood of neutrinos released by the events going on in the core — conditions in the shock are so extremely dense that even neutrinos are stopped in it, and help to push the outermost layers of the star away. But only relatively small amounts of the very heavy elements can ever be produced, because the conditions under which they can be made never last long. The total mass of all the elements heavier than iron put together is only 1 percent of the total mass of all the elements from lithium to iron — and the total mass of *all* the "metals" is less than 2 percent of the mass of hydrogen and helium around. The ultimate effect of the supernova is that the blast from the core of the dying star pushes ten or more solar masses of material away into space. This time, there is very little iron in the expanding cloud of material, because it was all left behind in the core; but there could be one or two solar masses of oxygen in the ejected material, along with those traces of very heavy elements and other bits and pieces.

This brief overview of what we know — or what we think we know — about how the chemical elements are made may give the impression that everything is cut and tried. That's true up to a point — the broad outlines are indeed clear. But this is really a work in progress, and I wouldn't want you to go away with the idea that there is nothing left to discover about the origin of the elements. In order to understand the processes properly, you would need a full understanding of *all* the nuclear interactions involved — an understanding which can only come from studying the behavior of the nuclei in the laboratory. That's a tall order, both because of the numbers of interactions involved and because of the very short lifetimes of many of the nuclei that take part in those interactions. There are 116 known elements, which come in a total of about 300 naturally occurring varieties (300 isotopes) on Earth. But theory suggests that about *six thousand* isotopes could exist in principle, some of them very short-lived; any of them could be involved in interactions inside stars, but more than half of them have still to be detected in accelerator experiments.

Such experiments, searching for "new" isotopes, measuring their properties and the way they interact with other nuclei, continue at laboratories around the world; there is a dedicated project at Michigan State University, for example, and plans for a $1 billion Rare Isotope Accelerator to be built in the second decade of the twenty-first century. In reality, our understanding of the origin of the elements is still pretty basic, and over the next ten or twenty years we can expect real progress in understanding exactly what goes on inside stars and supernovas, how the chemical elements we are composed of were manufactured, and why they are present in the Universe in the exact proportions we observe. But heavy elements are not the only by-products of supernova explosions.

In the second kind of supernova (known as Type II; the simpler ones are Type I), the core itself is left behind at the site of the explosion. The ball of stuff definitely has more mass than the Chandrasekhar limit, and if it has less than about three solar masses it has one final possible resting place. It will stabilize as a ball of neutrons (essentially, a single huge "atomic" nucleus) with more mass than the Sun packed into a sphere about 10 kilometers across. Many such stars have been identified by the radio noise they emit; they are known as pulsars and are often found, as you would expect, at the hearts of the expanding clouds of debris left behind by supernova explosions. I have a fondness for pulsars, since the first important piece of research I ever did, as a Ph.D. student, was to show that pulsars could not be white dwarfs and must therefore, by a process of elimination, be neutron stars. But if the core has more than about three solar masses, there is no way it can hold itself up against its own gravitational pull, and it will collapse entirely toward a point (a singularity mirroring the singularity at the birth of the Universe), shutting itself off from the outside world as its gravitational field becomes too strong to allow even light to escape.

The same questions surround the black hole singularity as those concerning the singularity at the birth of the Universe. Does everything really collapse to a point with zero volume, or does some property of space and time (membranes?) prevent this? Nobody knows, since by definition we cannot see inside a black hole; but the same ideas that are leading to a new understanding of the birth of space and time in the Big Bang will also lead to a better understanding of the "death" of space and time under such conditions. Black holes are often linked with death and destruction in popular accounts and fiction, portrayed as the ultimate doomsday phenomena, roaming the Galaxy and swallowing everything they encounter; but it's worth remembering that the same processes that make black holes also seed the Universe with the chemical elements required for life. Our existence is intimately bound up with the existence of black holes.

Although we have pointed it out before, it's worth a small aside to mention just how quickly the processes we have just described happen in massive stars. The more massive a star is, the more furiously it has to burn its nuclear fuel to hold itself up against the inward tug of gravity. And at each step up the fusion chain—from hydrogen burning to helium burning to simple carbon burning and so on—the energy released in each interaction is less, so the fuel lasts for a shorter time. Our Sun is 4.5 billion years old and still only halfway through its time as a hydrogen-burning star. But in a star which starts out with seventeen or eighteen solar masses, hydrogen burning lasts for just a few million years, helium burning about a million years, carbon burning a mere twelve thousand years, neon and oxygen burning about ten years, and silicon fusion just a few days. The most recent developments in our understanding of the origin of the chemical elements have, however, come from the opposite end of the scale — from stars smaller than the Sun, which have such immense lifetimes that they are still around today even though they formed when the Universe was young.

Everything we have told you about stellar nucleosynthesis so far refers to the kinds of stars we see in profusion in the Milky Way Galaxy today. All these stars are enriched with elements heavier than helium, and so cannot be made of the basic baryonic material that emerged from the Big Bang. There must have been at least one generation of "original" stars, which manufactured, for example, the carbon that participates in the CNO cycle in stars somewhat more massive than our Sun today. The common stars of the Milky Way and similar galaxies come in two basic families, known as populations. Population I stars are stars like our Sun, mostly in the disk of the galaxy, which contain the greatest proportion of heavy elements. They have been produced from the material processed in several previous generations of stars, and because the clouds of interstellar material from which they formed were relatively rich in heavy elements, they are the most likely stars to be associated with planets and life (as we will discuss in more detail in the next chapter). The second population of stars, called Population II, are found chiefly in a spherical halo that surrounds the disks of the Milky Way and similar galaxies; they are older stars which, because they formed when the Universe was younger and less material had been processed inside previous stars, are less enriched with heavy elements than Population I stars. There is very little chance of finding a rocky, Earth-like planet associated with a Population II star. But spectroscopy shows that the outer layers of even these stars, the cooler regions where nucleosynthesis has not been going on, still contain traces of heavy elements, so these are not the first stars to have formed.

By a logical extension of the naming system that gave us Population I and Population II, the first stars in the Universe, made solely of hydrogen and helium, are referred to as Population III, even though they have not yet been seen. They must have been there to make the traces of heavy elements seen even in Population II stars, and they started the process which led to the formation of stars like the Sun, planets like Earth, and people like ourselves. But it is much harder to make stars out of hydrogen and helium alone than to make them out of clouds of hydrogen and helium laced with a trace of heavy elements. The problem is that as a cloud of gas collapses under its own gravitational pull, it gets hot inside, and this heat tends to blow the cloud apart before it can become compact enough to form a star.

If there are traces of things like carbon and oxygen around, however, molecules of compounds such as carbon monoxide and water vapor can form. These molecules get hot along with the rest of the material as the gas cloud collapses, but they are very good at radiating heat away in the form of infrared energy. This allows the cloud to radiate excess heat away and continue to collapse until it forms stars like the Sun. Without those atoms and molecules present, though, the collapse can occur only if the cloud has a large mass, at least several tens of solar masses. If it is that big, the cloud will quickly collapse under its own weight, producing the temperatures required for nuclear fusion in its interior and exploding as a supernova in a blast which disrupts the star completely and scatters heavy elements into the interstellar medium. The death throes of such stars produced such huge explosions that they are detectable today as burst of gamma rays coming from the edge of the observable Universe — the most distant of these outbursts yet detected is estimated to have come from a redshift of 6.3, corresponding to a time when the Universe was less than a billion years old.

Until very recently, it seemed that all of the original Population III stars probably had masses well in excess of a hundred times the mass of the Sun and ran through this cycle in much less than a million years. This is good for making the heavy elements seen in Population II stars, and for producing gamma ray bursts, but bad for the prospects of astronomers trying to find Population III stars that have survived from the era just after recombination. Only small stars burn their fuel slowly enough to have survived since then, and conventional wisdom had it that there were no small stars formed in that era.

Not for the first time, though, conventional wisdom proved wrong. In the early years of the twenty-first century, several small, faint stars which contain very little in the way of metals were identified in our Galaxy. They are not quite pure Population III stars; but they do seem to be "fossil" remnants from the dawn of time. Their existence has already helped astronomers to infer what kind of stars their Population III progenitors really were, and they seem likely to yield further insights into the way star formation began as more of them are investigated.

The new discoveries came as a result of a survey of a large area of the southern sky, carried out over an interval of ten years by an international team of astronomers using the best modern telescopes — an archetypal example of the way in which progress in science today depends on large collaborations using expensive hi-tech equipment rather than the isolated genius working alone in a lab.* The faint stars

---

*The survey was actually aimed primarily at finding quasars; the faint star discoveries were a bonus, which highlights one of the advantages of such big projects — the fact that they gather enormous amounts of data which can be used in many different ways.

revealed by this survey are composed almost entirely of hydrogen and helium, with typically less than one-two-hundred-thousandth of the proportion of "metals" found in the Sun. In this case the term *metals* really is apposite since the stars are indeed almost entirely devoid of iron but do contain traces of carbon and nitrogen. The inferred ages of these objects are above 13 billion years, which means that they formed within a billion years of the Big Bang and provide us with direct clues to the nature of the Universe at that time. Yet these are stars in our own Galaxy, no more than a few thousand light years away, providing insight into conditions that could only otherwise be seen at extremely high redshifts, taking advantage of the look-back time.

The first surprise was that such small stars (weighing in with about 80 percent of the mass of our Sun) could form with such a small trace of carbon and other heavy elements to provide the infrared cooling mechanism required in order for the clouds from which they form to collapse (see Chapter 8); the second puzzle was where the trace of carbon and nitrogen could have come from, without any iron being produced by the original Population III stars. The best solution to have emerged so far comes from two researchers at the University of Tokyo, who calculated in detail the life cycle of Population III stars which start out with masses in the range from 20 to 130 times the mass of the Sun. They found a particularly good match with the element abundances seen in the extremely metal-deficient old stars if the progenitor stars have a mass of about twenty-five solar masses; but the observed element abundances cannot be matched by progenitor stars with masses in the range from 130 to 300 times the mass of our Sun.

The key feature of the life cycle of the Population III stars with masses a few tens of times the mass of the Sun is that they are not completely disrupted at the end of their lives. Although they do explode as Type II supernovas, and the outer layers, rich in carbon and nitrogen, are ejected into space, the explosion is not violent enough to disrupt the iron core of the star. Instead, the material of the core, rich in iron and other heavy elements, falls back upon itself to make a remnant with a mass between three and ten solar masses. This is above the limit of stability for a neutron star, so the remnant must be a black hole. Crucially, this is not just a matter of theory and computer modeling of the life of a star; there is a known class of Type II supernovas, called "faint supernovas," whose observed behavior closely matches the predictions of the computer simulations.

This makes the model doubly attractive. As well as giving exactly the kind of enrichment of the interstellar medium needed to explain the element abundances in the extremely metal-poor stars, it provides the first black holes, which, as we have seen, are themselves important in encouraging stars to cluster together as the Universe expands. There would have been copious numbers of these black holes in the early Universe, which could have merged and grown to become the supermassive objects now detected at the centers of galaxies.

So we can trace the origin of the chemical elements all the way back to the formation of the first stars, shortly after the time of recombination. The history of star formation is itself written in the composition of the stars, with the oldest stars containing the smallest fraction of heavy elements and the youngest stars containing the richest mix of elements. The story begins with the mixture of hydrogen and helium, plus traces of deuterium and lithium, that emerged from the Big Bang. The next few million years were dominated by stars with a few tens of solar masses, and lifetimes of less than a million years. They provided the raw material for the next generation of stars, and the smallest of these next-generation stars survive today as the extremely metal-poor stars. But larger second-generation stars (Population 2.5?), with masses of perhaps eight to ten times the mass of our Sun and lifetimes of only tens of millions of years, dominated the scene for about 30 million to 100 million years after the Big Bang, building up heavier elements such as barium and europium and scattering them through the interstellar medium when they went supernova at the end of their lives. The enriched material provided by this generation of stars allowed stars with masses of only three to seven times the mass of the Sun to form in profusion, and it was these stars that began to produce and distribute across space the kind of mixture of heavy elements that is seen today in the Sun and other stars of its generation.

Because smaller stars live longer, the era dominated by these stars lasted from about a hundred million years after the Big Bang to a billion years after the Big Bang. It was only after this era had enriched the material between the stars still further that the kind of stars which spread iron across space, in the way we described earlier, were able to form. But by three or four billion years after the Big Bang, roughly ten billion years ago, galaxies like the Milky Way were already in existence, with their two distinct populations of stars, and with star formation proceeding in the disk of the Milky Way more or less as it does today. As time passes, the interstellar medium and later generations of stars continue to be increasingly enriched with heavy elements; but the process has been continuing in essentially the same way, a quasi-steady state, for all that time. It is against that background that we can take a look at how the Sun and its family of planets, the Solar System, formed, when the Milky Way was about half its present age.

8

Where Did the Solar System Come From?

The Ancients thought that the stars were eternal and unchanging. From a modern perspective, even if you knew that the Universe had a beginning, you might guess that all the stars were born shortly after the beginning, and have been around ever since. But as we have explained, we know that there have been several generations of stars since the Big Bang. We also know that stars are still being born in the Milky Way and other galaxies today, and by studying the sites of starbirth we can begin to understand how the Sun and its family of planets formed. These studies have dramatically changed our ideas about starbirth over the past few years, as improved technology and new observing instruments have enabled astronomers to probe into the hearts of the clouds of dust and gas in which stars form.

It was the link between young stars and clouds of gas and dust in space that first provided the clue to where stars and planets come from. The life span of a star depends on how quickly it is burning its fuel (which means how bright it is) and how much fuel it has to burn. More massive stars have more fuel to burn, but they have to burn it much faster to hold themselves up against the pull of gravity. So the shortest-lived stars are both big and bright. Because such stars have such short lifetimes, when we see them we know that we are seeing them close to the sites in which they were born — and the biggest, brightest stars in our Galaxy are all associated with clouds of dust and gas. One of the nearest such aggregations of material is located in the bright constellation Orion. The famous Orion Nebula is just part of a much larger region of clouds of gas and dust, and other interesting astronomical objects, sometimes known as a "gas-dust complex." Some of the most famous pictures taken with the Hubble Space Telescope show young stars embedded in the clouds of the Orion complex, with the clouds themselves being etched away by the radiation from the stars.

Part of the reason why astronomers are now confident that they understand at least the broad picture of the processes of star formation is that in star-forming regions like Orion there are thousands of young stars to study. One of the most intensively studied of these systems is the Orion Nebula Cluster, which lies about 450 parsecs (some 1,500 light years) away from us. With a look-back time of 1,500 years, the light now reaching us from this cluster set out on its journey about the time that Muhammad was teaching on Earth. The central, most populous heart of the cluster has a radius of a fifth of a parsec, with a density equivalent to 20,000 stars per cubic parsec; this is embedded in a region of lower stellar density extending out to a radius of 2 parsecs and containing at least 2,200 stars.

It is also now possible to measure the ages of some of these stars with reasonable accuracy, rather than just saying, "They are bright, so they must be young." Although the processes of nuclear fusion that go on inside stars steadily build up more complex nuclei, so that in general newly formed stars have a richer mixture of elements heavier than helium than do stars which formed long ago, there is one exception to the rule. Lithium, element number 3 in the periodic table, is not manufactured inside stars at all, and all the lithium in the Universe today is left over from Big Bang nucleosynthesis. Worse (as far as lithium is concerned), stars actually "burn" lithium in some of the nuclear interactions that go on inside them. So in each generation of stars, there is *less* lithium available than there was to the previous generation. Which means that the stars which contain the least lithium are the ones that have just formed. This way of dating stars is astonishingly precise; early in the twenty-first century astronomers using the technique found that more than twenty of the stars in the Orion Nebula, all of which have masses comparable to that of our Sun, are less than ten million years old, and the youngest of them were born only about a million years ago. This is the best and most direct evidence that young stars are indeed associated with clouds of gas and dust in the Milky Way Galaxy; the youngest ages, around a million years, also match the ages of a class of stars known as T Tauri stars, inferred from a comparison of their observed properties with theoretical models of young stars.

The natural assumption, from studies of many such complexes, is that stars are born in the middle of such clouds, out of material pulled together by gravity, but the clouds are then dispersed, partly because of the outward pressure of light and other radiation from the young stars. The biggest and brightest stars burn out soon afterward, but smaller, longer-lived stars (like our Sun) are left to wander around the Milky Way for thousands of millions of years, losing all contact with the region in which they were born.

This broad outline of starbirth has been clear for nearly a hundred years, although the details were obscure until very recently. Indeed, the link between clouds of diffuse material and the origin of stars is such a natural one to make that back in the seventeenth century Isaac Newton wrote to Richard Bentley, who had asked for information about the Universe:

It seems to me, that if the matter of our sun and planets, and all the matter of the universe, were evenly scattered throughout all the heavens, and every particle had an innate gravity towards all the rest, and the whole space throughout which this matter was scattered, was finite, the matter on the outside of this space would by its gravity tend towards all the matter on the inside, and by consequence fall down into the middle of the whole space, and there compose one great spherical mass. But if the matter were evenly disposed throughout an infinite space, it could never convene into one mass; but some of it would convene into one mass and some into another, so as to make an infinite number of great masses, scattered great distances from one to another throughout all that infinite space. And thus might the sun and fixed stars be formed, supposing the matter were of a lucid nature. [Jeans, *Astronomy and Cosmogony*]*

Gravity does indeed pull clumps of matter within interstellar clouds together to make new stars. But this must be a very inefficient process. The Milky Way has been around for well over ten billion years, yet there are still clouds of gas and dust between the stars, and still some star-forming activity today. Why didn't all the material con-

---

*Newton's comments seem even more prescient if you apply them to galaxies rather than stars; but, of course, he did not know that the Milky Way is just one galaxy among hundreds of billions in a vastly bigger Universe.

dense into stars long ago? Because there is a significant difference between the kind of static cloud of material that Newton envisaged and the dynamic state of the material in the Milky Way Galaxy. It is certainly true that a static cloud of gas and dust would collapse under its own weight, at least to the point where it became hot enough inside to hold itself up. But all the material in the Milky Way, including stars themselves and the material from which stars are made, is in motion. If the Earth were at the same distance from the Sun but stationary relative to the Sun, it would immediately begin to fall straight into the Sun; it doesn't because it is in orbit, moving around the Sun. The material the Milky Way Galaxy is made of is also in orbit, around the center of the Milky Way, and the interstellar clouds themselves rotate slowly around their own centers, with other random motions superimposed on this more or less circular motion. There are "winds," comparable to those in the atmosphere of the Earth, that move gas around within the clouds; and there are magnetic fields which thread the clouds and help prevent them collapsing.

With all of this going on, the surprise is that stars are formed at all; indeed, astronomers estimate that only a few times as much material as there is in our Sun (a few solar masses) is converted into new stars each year in our Galaxy. This is roughly in balance with the amount of material ejected back into space by old stars when they die. One implication of this is that many stars must indeed have been born in a short span of time during the process which made the Milky Way all those billions of years ago, before it settled into its present steady state. Such events, known as "starbursts," are still seen in other galaxies today; but we will not discuss them further, because that is not the way in which our Solar System formed, a mere five billion years ago, when the Milky Way had already been settled into its present way of life for several billion years.

The random motion of the clouds, and of the gas within the clouds, can be studied by spectroscopy, with the aid of the Doppler effect. Such studies can also reveal other details of the conditions inside the clouds, such as their density and temperature. In the "empty" space between the stars, there is on average about one atom in every 15 cubic centimeters (and by far the majority of those atoms, of course, are hydrogen). In an average sort of interstellar cloud, there would be about 10,000 atoms in the same volume, and the cloud itself might extend over a distance of thirty or forty light years,* roughly four times the distance from the Sun to its nearest similar stellar neighbor, Tau Ceti. Although clouds in which bright stars are embedded may be as hot as 10,000 °C, one of the most important features of the clouds in which the collapse to form new stars begins is that they are extremely cold, less than 10 degrees above the absolute zero of temperature (less than 10 K, or below *minus* 263 °C).

Most of the clouds in our Galaxy are not contracting to form stars. They are more or less in equilibrium, supported in particular by magnetic fields and rotation. If a star with the mass of the Sun formed directly from the collapse of a slowly rotating gas cloud containing the same mass but spread out with the density of an interstellar cloud, it would spin faster and faster as it shrank, like a spinning ice skater pulling in his or her arms. By the time it was Sun-sized, the equator of the object would be spinning at about 80 percent of the speed of light, which is obviously ridiculous. So one of the key processes in star formation is getting rid of this excess spin, or angular momentum. The best way to do this is to start with a lot more mass and throw some of it away into space, carrying angular momentum with it, as the

---

*In the units more usually used by astronomers, some 10 to 13 parsecs; 1 parsec is 3.26 light years.

central region collapses; this is seen to be happening in star-forming regions, but the details of just how the excess material gets ejected are still not clear. Another solution to the problem is if two (or more) stars form from the same collapsing cloud; then much of the angular momentum of the cloud is converted into the orbital motion of the stars around each other. Significantly, about two-thirds of all stars are in binary pairs or more complicated systems. On a much smaller scale, the planets of the Solar System, orbiting round the Sun, also carry angular momentum stored up from the collapse of the cloud in which the Solar System formed. A star without planets is likely to be spinning faster than a star with planets.

When I was a student, we were taught that stars formed from a collapsing cloud through a fairly steady ("quasi-static") process of collapse and fragmentation. A cloud containing a thousand solar masses of material might be expected to produce about a thousand stars, contracting slowly at first until it became unstable, when different parts of the cloud would begin to collapse under their own self-gravity; then those pieces of the cloud would fragment and the smaller pieces would collapse in their turn, with the process repeating several times. The process of collapse and fragmentation continued fairly steadily, with gravitational energy being converted into heat along the way, until the fragments became hot enough to shine as protostars. According to this old-fashioned picture, all stars form in association with other stars, but once they have formed they follow their own orbits around the Galaxy, so that within a few hundred million years they are spread far apart from one another, and there is no way to trace their common origin.

But now we think that star formation is a much more violent, rough-and-tumble process. It is still clear that stars do form together in this way, but not as a result of the smooth and gradual collapse of a cloud of gas and dust. The change in our thinking has come as a result of better observations and better theoretical models of what goes on inside those clouds. Because the clouds contain so much dust, their cores are only visible in the infrared part of the spectrum — infrared light penetrates the dust.* But infrared radiation is largely blocked by water vapor in the Earth's atmosphere, so it has not been possible to peer into the centers of star-forming regions until the development of satellites to carry infrared telescopes above the atmosphere, and of infrared telescopes sited on high mountaintops, with most of the atmosphere (and essentially all the water vapor) below them. Some of the most important of these new observations have come, for example, from the James Clerk Maxwell Telescope (JCMT) at the Mauna Kea Observatory in Hawaii; one of the key instruments attached to the telescope is known to astronomers, with their fondness for acronyms, as SCUBA (the Sub-millimeter Common-User Bolometer Array). At the same time that the new observations have been coming in, since the mid-1990s, improvements in computer power and speed have made it possible to simulate what goes on inside the clouds, just as the same improvements in computers have revolutionized our understanding of how structure developed in the expanding Universe after the Big Bang.

There are other observations, which go back to the middle of the twentieth century but are still being improved upon, that provide insight into the nature of gas-dust complexes and how they are able to give birth to stars. Among them is the discovery that such systems contain not only atoms but molecules as well. These molecules are identified by spectroscopy at radio wavelengths, and among the first

---

*Even red light penetrates dust better than light with shorter wavelengths, which is why sunsets, seen through the dusty lower layers of the Earth's atmosphere, look red.

to be found were CH, CN, and OH. There are also more complex molecules, including H_2HCO, CH_3HCO, and CH_3CN. Molecular hydrogen (H_2) has also been detected (all of which gives such complexes another name—giant molecular clouds). The fact that such molecules exist in the clouds reveals a great deal about the conditions there. For example, hydrogen atoms that collide with one another in empty space would simply bounce off each other; they can only manage to stick together to make molecules if they interact on the surface of a grain of dust. So the density of molecular hydrogen in the clouds tells us that there must be at least a hundred dust grains, comparable in size to the particles in cigarette smoke, in every cubic centimeter of such a cloud. The dust grains also do another job. If they were not there, ultraviolet light penetrating the cloud from nearby stars would break the hydrogen molecules apart; but the dust shields the hydrogen from the ultraviolet to such good effect that in a dense cloud such as the Orion Nebula there must be as many as ten million hydrogen molecules in every cubic centimeter.

The molecular gas in the Milky Way is concentrated in large complexes by the effects of magnetic fields and tides in the rotating disk of the galaxy. Large complexes may be as much as a thousand parsecs across, and contain ten million solar masses of material, but an individual giant molecular cloud will be no more than 100 parsecs across and have a mass of up to a million solar masses. The average (median) size is abut 20 parsecs, and the average (median) mass is about 350,000 solar masses. This material can also be swept up into a cloud by the explosion of a supernova, which sends shock waves rippling through the interstellar medium; there is direct evidence, from the abundances of rare isotopes found in meteorite samples, that the cloud from which our Solar System formed was hit by just such a shock wave about a million years before the Sun was born. The distribution of matter inside such a cloud looks remarkably like the distribution of matter seen in simulations of the formation of structure in the early Universe, with sheets and filaments of gas and knotty clumps of denser material where the filaments meet; gas flows along the filaments and accumulates at certain points, especially where filaments cross. This pattern, like beads on a necklace, is repeated on smaller scales within the cloud, so that apart from its size the kind of pattern of gas within a subunit is the same as the kind of pattern within the "parent" cloud; this hierarchical structure is approximately fractal. Now that they can be studied in detail, the very appearance of the clouds, with irregular and filamentary shapes that actually look windblown (like a close-up of a cloud in the atmosphere of the Earth seen from an aircraft), shows that they are far from being in equilibrium.

All of this fits a relatively new picture of what goes on inside molecular clouds, a model dominated by the effects of turbulence in the gas. Turbulent flows, rather than magnetic fields, are now seen as the key to preventing the rapid collapse of the clouds, and collisions between flows of gas moving with supersonic speeds (which means faster than about 200 meters per second, under these conditions) produce shock waves in which parts of the cloud can begin to collapse to form stars. Gravity and turbulence are equally important in determining the structure of these clouds and how they evolve; stars only form where gravity dominates locally. It is also clear that the clouds themselves are unstable, short-lived phenomena, by the timescales of the Milky Way. But just what causes the high-speed turbulent flows remains something of a mystery, so although we know the turbulence exists, when it comes to understanding it, and the details of star formation, we are back in the realm of what we *think* we know.

One thing we do know for sure is that the clouds can only collapse because they are very cold. The same dust that prevents ultraviolet radiation from outside from disrupting the hydrogen molecules shields the hearts of the clouds from sources of energy, while the molecules in the cloud, in particular carbon monoxide (CO), radiate infrared energy and cool the cloud. (Dust on the fringes of a cloud gets warm as it absorbs energy from outside, but this doesn't matter to the cold core of the cloud.) It is only because they are cooled below 10 K that cloud fragments as small as one solar mass can collapse at all; if they were any warmer, the outward pressure resulting from their heat energy would be enough to prevent gravity from making them collapse.

Although lots of interesting things go on in the giant molecular clouds on different length scales and different timescales, we are especially interested in where stars like the Sun (and planets like the Earth) come from, so we will only look in detail at what happens on this relatively modest scale. The most important thing, which is clear from both observations of the stars associated with such clouds and the computer simulations, is that stars do not form in isolation. By far the majority of stars form in association with at least one or two companions, and it is possible that all stars form in this way. So isolated stars like our Sun have most likely been ejected from such systems at an early stage in their lives.

This is actually an easy and natural process, involving nothing more complicated than gravity and the laws of orbital mechanics worked out by Isaac Newton three and a half centuries ago. If three stars with roughly the same mass are in orbit around one another, it is almost inevitable that sooner rather than later a kind of gravitational slingshot effect will send one of the stars hurtling away into space, carrying angular momentum with it, while the other two stars move into a closer orbit around each other. The same sort of thing can happen if there are more than three stars in the group to start with, but when there are just two stars left they remain as a binary pair locked in a gravitational embrace.

Partly from our new understanding of the dynamic nature of the giant molecular clouds, and partly from the clear observation that very few of these clouds are known which do not contain star-forming regions, it is clear that there is no significant pause between the formation of a giant molecular cloud and the onset of star formation. The cloud coalesces, stars form, the radiation from the hot young stars blows the cloud away, and within ten million years the whole process is over. A cloud forms, does its star-forming thing, and disperses all within about the time it would take a sound wave to travel from one side of the cloud to the other — what astronomers refer to as the "crossing time."

Remember, though, that one of the reasons why a cloud of gas and dust collapses to form stars is because it has been compressed by just the same kind of shock waves from supernovas and winds from young stars, spreading out through the interstellar medium, that disperse the local material surrounding those stars soon after they are born. Intriguingly, evidence from the Galaxy as a whole suggests that this is a self-regulating process that can operate more or less in a steady state for many billions of years. If a lot of stars form in one generation, the gas and dust gets blown away and diffuses, which makes it harder to form new stars. If only a few stars form, the gas is not so widely dispersed, and it is easy for stars to form in the next generation. So if ever the process moves toward one extreme or the other, it tends to shift back toward the long-term average in the next generation.

Hardly surprisingly, the youngest stars in a molecular cloud are found in the regions of highest density in the cloud; but it is still an open question to what extent this clumpy structure (often with a hierarchy of clumps inside clumps inside clumps) is a result of gravity overwhelming pressure forces when the density is great enough, and to what extent the shock waves associated with turbulent flows, which can squeeze local regions to a density a hundred times greater than their immediate surroundings, play a part. The calculations suggest, though, that turbulent compression can result in the creation of prestellar cores, regions of higher-than-average density on the brink of gravitational collapse. Such cores are calculated to have an internal temperature of about 10 K and may be about 0.06 of a parsec (about a fifth of a light year) across, containing about 70 percent as much mass as our Sun. Cores seen in molecular clouds have very similar properties to these, making turbulent compression the front-runner among the explanations for the formation of prestellar objects. Then gravity takes over — hindered by magnetic fields and the need to get rid of angular momentum.

But although close to half the mass of gas might get converted into stars in the most dense clumps within a cloud, over the whole cloud probably no more than a few percent of the material is turned into stars before the cloud disperses. As we focus in on the details of how an individual star does form, it's worth remembering that star-formation is not, overall, a very efficient process.

However a prestellar core forms, though, we know that such cores exist because we can see them, and we can use them as the starting points for a detailed explanation of how stars like the Sun form. Our understanding of what happens next depends almost entirely on computer simulations, and different models (for example, with different amounts of magnetic influence) make different predictions about how details of the collapse occur, and in particular how fast it happens. But all the simulations make the same intriguing prediction — that in the first stage of the formation of a star, only a tiny fraction of the mass of the collapsing cloud will reach the density necessary to turn it into a star. Instead of a whole cloud with the mass of the Sun shrinking down upon itself almost uniformly until it becomes an object like our Sun, right at the center of the cloud an object with only 0.001 solar masses becomes dense enough and hot enough to stabilize and soon afterward to begin to generate energy by nuclear fusion. Then, over a much longer period of time, the rest of the mass of the star is added by accretion, as material from the surrounding cloud falls onto the surface of the protostar.

Rotation and magnetism do not affect this process of forming a tiny initial protostar, although rotation is responsible for the collapsing cloud forming two or three such protostars orbiting around one another instead of a single core. A rotating cloud collapses to a disk of material around the protostar (one disk around each protostar), and according to some calculations an uneven distribution of matter in such a disk can cause fluctuations which carry some of the material outward, taking angular momentum away from the protostar, while some is driven inward to fuel the accretion. Any such wave with a trailing pattern (with the arms of the spiral bent back relative to the rotation) always transports angular momentum outward, leading some astronomers to describe accretion disks as machines for stripping gas of its angular momentum. Another possibility is that magnetic fields trapped by the protostar funnel some of the material falling in toward the star back out into space as jets emitted from the poles of the star, again getting rid of excess angular momentum; such jets are seen in many young stars, although this explanation of them is as yet no more than an educated guess.

The great beauty of all these calculations is that they tell us that whatever kind of cloud you start out with, and whatever kind of rotation, magnetic fields, and other properties it has, you always get the same kind of central compact object forming. So we don't have to worry about exactly what went on before by the time we get to the point where the central region begins to warm up. This happens when the density is high enough for the infrared radiation to be trapped instead of escaping from the core; the required density is about 10^{-13} grams per cubic centimeter, which corresponds to about 20 billion hydrogen molecules in every cubic centimeter. The rising pressure inside the core then brings the collapse to a halt when the central density has increased by a factor of about two thousand, with 40,000 billion hydrogen molecules in every cubic centimeter (about one-fifth of a billionth of a gram per cubic centimeter). This core has a mass roughly one-hundredth of a solar mass, and extends over a volume several times bigger in radius than the distance from the Earth to the Sun. But its stability is short-lived, because the temperature inside continues to rise; and when it reaches 2,000 K, hydrogen molecules are broken apart into their constituent atoms.

This alters the behavior of the gas and allows a second phase of collapse, which occurs in the same way as before to produce a new "inner" inner core. The collapse of this core is brought to a halt only when it becomes hot enough inside for electrons to be stripped from the hydrogen atoms to form an ionized plasma, which happens by the time the temperature reaches 10,000 K; but this time, the end of the collapse is permanent. This process of ionization is, of course, the reverse of the process of recombination that occurred in the Universe at large when it was a few hundred thousand years old. Just as the Universe became transparent at that time, because there were no longer large numbers of charged particles around for light to interact with, so when the prestellar core ionizes it becomes opaque, because electromagnetic radiation inside the core is bounced around between the charged particles. This seems as good a definition as any of the moment when the core becomes a star.

It is this inner core that forms the seed on which the star will grow — a protostellar core with a mass of no more than one-thousandth of the mass of the Sun, occupying a volume about the same as that of the Sun today, but steadily increasing its mass as more material falls onto it from outside (the radius increases only slightly, to a few solar radii, since the main effect of the accretion is to make the protostar more dense). All the gas of the first core falls onto the protostar within about ten years, bringing its mass up to about 0.01 solar masses, but that still leaves by far the bulk of the mass of the original collapsing cloud to continue falling onto the protostar. As we have mentioned, the formation of a disk of material around the central star is an essential feature of this process (except for the unrealistic case of a nonrotating, nonmagnetic object), and since the advent of the Hubble Space Telescope late in the twentieth century dusty disks of material have been seen around many young stars. This has obvious implications for the formation of planets, and we will return to that topic shortly.

The size of the star that grows from the prestellar seed depends on the amount of material available for it to accrete, not on the size of the core, since all cores start out with much the same mass. Crucially, though, the way the fragmentation and collapse occurs means that each star (or each group of two or three stars) forms from a fragment of the molecular cloud that has already become cut off from its surroundings; so there is a strictly limited amount of matter available to accrete, and the final mass of a star depends on the size of the fragment. In the case of a star like the Sun, this means that about 99 percent of its mass is acquired by accretion.

When the mass of the core has reached about a fifth of the mass of the Sun, it has become hot enough inside for nuclear fusion to begin; but this is not the proton-proton chain that provides the energy that keeps the Sun shining today. The first fusion process that occurs in such a core involves deuterium, the heavy version of hydrogen in which each nucleus consists of a proton and a neutron bound together by the strong force. When accretion finishes, in the case of a star with one solar mass the young star will have a radius about four times that of the Sun today, and will gradually shrink as it settles down into its life as a mature star — what astronomers call a "main sequence" star. During this shrinking, the energy which keeps the star shining mostly comes from the gravitational energy released as it shrinks; it is only when it gets hot enough inside (about 15 million K) for the proton-proton reaction to begin that nuclear energy takes over and stops the star from shrinking any further. But before the star settles down, whatever its mass, it always goes through a phase in which the material inside it is completely mixed up by convection. So the proportion of elements we see in the surface layers of any main sequence star today (including the Sun) is an accurate guide to the mixture of elements in the cloud from which the star formed. This is not affected by the processes going on deep inside the star which convert hydrogen into helium, or (in the case of some stars) helium into carbon, because stars on the main sequence are not fully convective, and the material in the core is not dredged up to the surface.

Astronomers describe the different phases in the accretion process in terms of four "classes," defined arbitrarily but giving a rough idea of how long each step in the process takes. The "starless" phase of the cloud collapse and early development of the core until it becomes opaque takes about a million years. Class 0 corresponds to an early phase of rapid accretion onto the core, which lasts for a few tens of thousands of years and sees at least half of the final mass accreted; Class I is the longest accretion phase, adding most of the remaining mass but at a slower rate, and lasts a few hundred thousand years; Class II corresponds to the emergence of a young T Tauri star still shrouded by dust, and lasts about a million years; Class III covers the period when a young star is no longer surrounded by dust and contracts onto the main sequence, which takes up to a few tens of millions of years.

The evidence for these timescales comes partly from modeling and partly from observation — for example, we see ten times as many Class I protostars around today as we do Class 0, but each Class I object must once have been a Class 0 itself, so the natural inference is that each star spends ten times longer in the Class I phase than it does in the Class 0 phase. Overall, this means that for a star which ends up with the same mass as our Sun, it takes about ten million years to evolve from a cloud of gas and dust on the brink of collapse into a main sequence star. A star which ends up with 15 solar masses takes only about a hundred thousand years to reach the same stage of its life.

As we have mentioned, by far the majority of stars (perhaps all of them) must form in multiple systems, if only because of the angular momentum problem, so that isolated stars like our Sun are wanderers that have been ejected by their former companions. It used to be thought that *multiple* might mean four, five, or more stars forming together, and some of the computer simulations also suggested that prestellar cores might fragment into many pieces. But an analysis carried out in 2005 by researchers at the University of Cardiff and the University of Bonn has shown that this is not the case. It turns out that it is very easy for mini-clusters containing many stars — "mini" because the globular clusters are much larger objects containing perhaps a million individual stars, which formed together when the Universe was young — to eject their members one at a time, which would lead to a far higher proportion of single stars observed in the Milky Way today than we actually see. By contrast, it is very difficult for such a system to eject pairs of stars bound together in a gravitational embrace. Indeed, the complete evaporation of such a "high-order" system to one binary and many single stars takes only about a hundred thousand years. In order to match the observed statistics, an individual cloud core must typically fragment to produce no more than three stars, although there may be occasional exceptions to this rule.

Typically, out of every hundred newly born star systems forty are triples and sixty are binaries. Of the forty triples, twenty-five are long-lived and relatively stable, while fifteen promptly eject one of their stars to provide fifteen binaries and fifteen single stars. This all happens in a star-forming region within about a hundred thousand years, leaving a ratio of twenty-five triples to seventy-five binaries to fifteen single stars. In the regions where the stars are born, such as the Orion Nebula today, close encounters between stars then disrupt more of the binary pairs, increasing the proportion of single stars in the Galaxy at large. Since each binary that is disrupted yields two single stars, just ten disruptions of this kind among the population we have described changes the ratio to 25:65:35, making single stars more common than triples.

Even so, by far the majority of stars around today are members of multiple systems, and at first sight it may seem a little odd that our Sun is one of the minority of single stars. But it may be that this is another example of our view of the Universe being biased by our own existence. The kind of dusty disks of material out of which planets like the Earth can form seem more likely to exist around isolated stars. In double or triple star systems, the gravitational influence of the other member(s) of the system would cause tidal effects which would disrupt the disks, and even if any planets did form they might end up in extreme orbits where they were alternately fried and frozen, or even swapped between the stars in the system. Life-forms like us can only exist on stable, long-lived planets in orbit around stable, long-lived stars; from this anthropic perspective, it is no surprise to find that our Sun is a solitary wanderer through space. Whatever the reason, though, it does mean that we do not have to worry about the complications caused by nearby companions when trying to understand just how the planets of the Solar System did form.

The essential features of a planet are that it is too small to get hot enough inside to "burn" nuclear fuel by fusion, and that it orbits around a star. But that still leaves a lot of possibilities to consider. We can narrow down the investigation further by restricting ourselves to planets which are, indeed, like the Earth. In the Solar System itself, there are four such small, rocky planets; they are the closest planets to the Sun and are known as Mercury, Venus, Earth, and Mars. There are also four large, gaseous planets orbiting farther out from the Sun — Jupiter, Saturn, Uranus, and Neptune. In addition there are various pieces of cosmic rubble, mostly lumps of ice or rock; one of these, Pluto, is generally classified as a planet for historical reasons. The distinction we want to discuss here, however, is between the rocky ("terrestrial") planets and the gas giants ("jupiters").

It used to be thought that both kinds of planet formed in the same way, by the accumulation of smaller pieces of material in the disk around a young star — often called the "bottom-up" scenario. In the case of both larger and smaller planets, the first object to form would be a rocky core. This was more or less as far as the process could go for the inner planets, the argument ran, because the heat of the young star would blow most of the gas away into the outer regions of the still-forming planetary system. But in the orbit of Jupiter, for example, a rocky lump with perhaps a dozen times the mass of the Earth would be able to accumulate gas and icy material by gravity, growing to its present size.

The big snag with this scenario, though, is that it would take many millions of years for the gas giants to grow in this way—in fact, if Uranus and Neptune developed in their present orbits, the simplest form of bottom-up process would have taken longer than the present age of the Solar System to have built them to their present sizes. That didn't matter so much in the days when the only planetary system we knew about was our own Solar System, and astronomers could still hope to find a better version of the bottom-up scenario to resolve the discrepancy. But over the past few years they have found well over a hundred other planetary systems. In almost every case, the planet orbiting another star has been detected because of its gravitational influence on the star, making it wobble to and fro as the planet orbits around the star. The wobble is too small to be observed directly, but it shows up through the Doppler effect in the spectrum of the star. The initially surprising feature of the discoveries is that the planets that have been detected in this way are large jupiters orbiting stars in orbits much closer to their parent star than Jupiter is to the Sun.

In one sense, it is not surprising that these early discoveries of planets outside the Solar System (extrasolar planets, or "exoplanets") have been dominated by such objects, since large planets in close orbits have the biggest influence on their parent stars and are the easiest to find using the technique. In 2005 astronomers were at last able to report that they had detected infrared light from a few of these planets directly, indicating that they have temperatures around 800 °C —the first light seen from any extrasolar planets. A little later the same year, astronomers photographed another exoplanet, orbiting a star 100 parsecs (225 light years) away in the constellation Hydra, at a distance of 8 billion kilometers (54 times the distance from the Earth to the Sun). Once again, though, this is a giant planet, about five times as massive as Jupiter. The smallest exoplanet yet discovered has a mass at least six times that of the Earth and zips round its parent star,

Gliese 876, once every 1.94 days, so it hardly counts as "Earth-like." If there are terrestrial planets in Earth-like orbits around other stars, it will need the next generation of observing instruments to find them. Given that they exist, it is no surprise that we have found hot jupiters; the surprise was that they exist. So why do they exist?

A natural explanation is that gas giant planets do not form by bottom-up accretion, but from the top down, as unstable clumps in the original disk of material around a young star. Such clumps can form anywhere in the disc, close to the star or farther out, and interactions between the planets and the disk could change the orbits of the giants, so that they move in or out from the orbits they formed in. This "migration" could explain how Uranus and Neptune got to be where they are now, having formed rather quickly a lot closer to the Sun just after the Sun itself formed. Simulations suggest that the top-down process may only have taken a few hundred years to form the gas giant planets.

These ideas themselves are still evolving, and in 2005 an international team of researchers from Brazil, France, and the United States put their heads together to come up with the best detailed simulation yet of what happened when the Solar System was young. Their jumping-off point was the evidence, known since the Apollo missions of the late 1960s and early 1970s brought back samples from the Moon, that many of the dark features we see on the face of the Moon were produced by an intense phase of bombardment by debris from space when the Solar System was about 700 million years old, sometime after the inner planets formed. This is known as the late heavy bombardment, or LHB. Combining this with the new understanding of how giant planets formed, the team found that all four of the giant planets must have formed in close proximity to one another, surrounded by a swirling mass of small objects, lumps of ice and rock known as planetisimals. Beyond the orbit of the outermost planet, there was still a disk of planetisimals left over from the earlier stages of the formation of the Solar System; the remnant of this disk, known as the Kuiper Belt, is still there today (the so-called "tenth planet" is a particularly large lump of debris in the Kuiper Belt). But if the new studies are correct, the present-day Kuiper Belt is really only a remnant of its former glory. Because of gravitational interactions, Jupiter slowly moved closer to the Sun while the three other giants moved outward, and some of the planetisimals were scattered in the same sort of way, some toward the Sun and some away.

At first, this was a gradual process. But some 700 million years after the Solar System formed, a dramatic change occurred when Saturn passed through an orbit with a period exactly twice as long as the period of the orbit of Jupiter around the Sun. This produced a rhythmic combination of gravitational tugging by these two planets on the other objects in the outer Solar System, a resonance rather like the resonance that occurs when a child on a swing pumps the swing higher and higher with small but carefully timed movements. The chief result of this process was to propel Uranus and Neptune out to their present orbits, with the radius of Neptune's orbit doubling suddenly, sending it out into the inner part of the Kuiper Belt and scattering huge numbers of planetisimals into the inner Solar System — there had to be a balance between what went out and what went in, in line with Newton's famous dictum "for every action there is an equal and opposite reaction." It was this flood of planetisimals that scarred the surface of the Moon, and presumably the Earth and other terrestrial planets as well, although the traces are less obvious here because the surface of our planet has since been reshaped by plate tectonics (continental drift) and erosion.

This hypothesis gained strength later in 2005, when analysis of the debris scattered into space by the collision of NASA's Deep Impact probe with comet Tempel-1 showed that the composition of the cometary material (in particular, the amount of ethane) corresponds to the kind of chemistry that would have occurred in such objects forming in the region of the Solar System now occupied by Uranus and Neptune. Tempel-1 seems to be one of the pieces from this early cometary belt that were scattered outward into deeper space by the migration of Uranus and Neptune.

Everything fits together neatly, which leaves the slower bottom-up process to account for the formation of terrestrial planets, and in particular the formation of the Earth itself, over a much longer interval in a Solar System where the giant planets were already in existence; this is where the planetisimals really come into the story. Leaving the gas giants out of the picture for now, we can pick up the story of how planets like the Earth formed from the time when the Sun was surrounded by a dusty disk of material. Before the 1990s, astronomers had assumed that the terrestrial planets must have formed from such a disk, but they had no direct evidence that dusty disks did exist around young stars. Since then, however, following a breakthrough discovery made with the Hubble Space Telescope, ever-improving observational techniques have revealed huge disks of material (now known as protoplanetary disks, or PPDs) around many of the young stars in our neighborhood. It is clear that these are an important feature of the way stars form. Since the disks are not seen around older stars, it is also clear that they have either been dispersed or turned into something else — planets.

These disks really are huge. One of the units of measurement used by astronomers is called the astronomical unit (or AU), which corresponds to the average distance of the Earth from the Sun (roughly 150 million kilometers). The best-studied PPD lies around a star called Beta Pictoris and extends over a diameter of 1,500 AU (some 225 billion kilometers). The Beta Pictoris system is estimated to be about 200 million years old, and the mass of material in the disk is about one and a half times the mass of our Sun. Most of this material will be lost from the system as it settles down. We can see this process at work in some cases, where jets of material are also seen, squirting out at right angles to the disk from the poles of the young star at the heart of the system; typically, such jets (visible in infrared light) may extend for a thousand astronomical units (150 billion kilometers). For comparison, the radius of the orbit of Neptune, the outermost giant in our Solar System, is just 30 AU. Significantly, the inner part of the disk in the Beta Pictoris system, a few score astronomical units across, is warped and distorted in just the way we would expect if there were planets orbiting within it. In other cases, there are actually gaps in the inner regions of the disks, about the same size across as our Solar System, suggesting that the material there has already been swept up into planets.

The Hubble Space Telescope alone has found hundreds of protoplanetary disks, but we'll mention just a few of them to give an indication of the kind of conditions that must have existed when our Solar System was young. Overall, the ages of the stars which still possess such disks range from a few tens of millions of years to a few hundred million years, and spectroscopic studies show that they have similar composition to our Sun (leaving out the hydrogen and helium and concentrating on the "metals"). Other properties of the radiation from the disks reveal that they are not the original "cigarette smoke" dust seen in interstellar clouds but have already been processed in some way, presumably by being formed into planetisimals which have then collided with one another and broken up to form a "second generation" of dust. The infrared radiation from some of the systems (including a disk around Vega, also known as Alpha Lyrae, the brightest star in the constellation Lyra) shows that the typical size of the dust grains is about 10 microns (ten-millionths of a meter). The mass of the dust itself is, of course, much less than the mass of the whole disk, because of the presence of a great deal of hydrogen gas which is steadily lost into space. For Beta Pictoris, there seems to be about a hundred times the mass of the Earth present in the form of dust.

The most exciting observations of PPDs are the ones which suggest the presence of planets in the disks. We have already mentioned that some disks are warped in the way they would be warped by the presence of planets and some have gaps just the right size to fit a system like our Solar System. One of the easiest systems of this kind available for us to study is the disk around the young star Fomalhaut, which is about 200 million years old and has roughly twice the mass of our Sun. This is tilted to the line of sight from Earth, which makes it straightforward to study the way in which the "central" star isn't central at all, but displaced to one side of the disk, suggesting the gravitational influence of several large planets. Fomalhaut is only 7.7 parsecs (25 light years) away, which makes it easy to study the system using instruments such as the Hubble Space Telescope. Within the overall dusty disk, there is a very sharply defined belt, or ring, with a width of 25 AU (25 times the distance from the Earth to the Sun) and a diameter of 266 AU, measuring from its sharply defined inner edge. This is nine times the diameter of the orbit of Neptune, the outermost large planet in our Solar System. The center of the ring is offset by 15 AU from the position of the star Fomalhaut itself — a displacement of 2.25 billion kilometers, half the radius of Neptune's orbit. This is hardly a subtle effect. The offset of the ring and its sharp inner edge show that there are planets orbiting closer to the star and sweeping up material from the disk; the ring itself may represent the early stages of the formation of Fomalhaut's equivalent of the Solar System's Kuiper Belt, the reservoir of icy material left over from the formation of our own planetary system.

Another clue to the presence of planets comes from the fact that all the dust in these disks seems to be relatively cool. Friction in the disk must inevitably make some of the dust grains drift inward, toward the central star, where they ought to become hot and radiate accordingly. The fact that this hot dust is not seen implies in some cases (such as Alpha Lyrae) that something is gobbling up the grains as they drift inward. It is hard to see what that "something" could be except planets.

In Beta Pictoris, the archetype for such systems, the warping of the disk can be explained if there is an object orbiting the central star at distance between 1 and 20 AU, with a mass between six thousand and six times the mass of the Earth. In addition, the thickness of this particular disk suggests that there must be solid objects at least a thousand kilometers across orbiting within it and stirring it up, otherwise it would have settled into a much thinner configuration, more like the rings of Saturn. The successor to the Hubble Space Telescope (sometimes called the Next Generation Space Telescope or NGST, but officially the James Webb Space Telescope and due for launch in 2011) should be able to detect gaps in the disk swept out by planets like Jupiter. But this will only be the icing on the cake, since we already know jupiters exist and we already know PPDs exist. We also know that planetisimals exist in these dusty disks, since the spectroscopic properties of the dust grains show that they are second-generation particles. It is easy to see how a collection of icy rocks can be tugged together by gravity to form a planet like the Earth. So the question of how planets like the Earth form boils down to the question of how the cigarette-smoke-sized particles of dust in interstellar clouds stick together in the first place to make planetisimals.

The crucial word there is *stick*. When tiny dust grains orbiting in the vacuum (or near-vacuum) of space collide with one another, they tend to bounce off, rather than sticking together. Until recently, astronomers had argued largely from wishful thinking, suggesting in a vague sort of way that dust particles might stick if one crept up behind another in very nearly the same orbit, and just bumped into it gently. But there is another factor which has to be taken into account. One of the most common compounds likely to be present in the molecular cloud from which a planetary system forms is water. Hydrogen is overwhelmingly the most common element present, and oxygen, although far less common than hydrogen, is the most abundant "metal," and third most common element after hydrogen and helium. Since hydrogen and oxygen eagerly combine with each other to make water, there had to be a lot of water vapor around in the cloud from which the planets formed. But there wasn't any liquid water. In the near-vacuum of space, where the micron-sized dust grains were no warmer than a few tens of Kelvin, the water vapor would condense onto the grains directly as ice. Studies carried out in laboratories on Earth where these conditions can be simulated show that when this happens, the molecules of water, which have positive electric charge at one end and negative electric charge at the other end, line up so that the whole sheath of ice around the grain is electrically polarized. This produces electrical forces that can make the icy grains stick together like the tiny bar magnets you find in some executive desktop toys.

The ice must have differed from the kind of ice we put in our drinks in another way. Because it was condensing from the vapor state straight onto the tiny grains, it formed a structure more like snowflakes than ice cubes. So each tiny grain of solid material (mostly carbon and silicon compounds) would have been surrounded by a fluffy outer layer, like a shock absorber, which would cushion the impact when the grains collided with one another. The collisions would occur with fluffy ice reducing the strength of any bounce, and the electric forces would be enough to do the rest. In tests carried out at the Pacific Northwest National Laboratory in Washington tiny hard ceramic balls (one-sixteenth of an inch [0.2 cm] in diameter) dropped onto ordinary ice in a vacuum chamber bounced back to 48 percent of the height they were dropped from; but the same balls dropped onto ice deposited from water vapor at a temperature of 40 K bounced back to only 8 percent of the initial height.

The "fluffy ice" effect is particularly good at growing planetisimals in the colder, outer parts of a young planetary system where the protoplanetary disks are seen today. In the hotter, inner regions where planets like the Earth formed, similar electrical effects would have occurred involving silicate grains. Either way, once the original grains had grown to a modest size they would begin to attract each another gravitationally, building up objects a kilometer or more across within about a hundred thousand years. It is the collisions between these (and larger) objects that produce the second-generation dust seen in PPDs.

Building rocky planets like the Earth out of planetisimals a kilometer or more across with the aid of gravity simply takes time — perhaps as long as 50 million years, but an eyeblink compared with the age of the Solar System. There is very little mystery about this stage of the process, and no need to elaborate on it here. But it does lead us to the biggest mystery of all — how did life emerge on Earth, and is there life elsewhere in the Universe?

9

Where Did Life Originate?

Everybody knows what life is; but there is no completely satisfactory dictionary or textbook definition of *life*. A working definition that will do for our present purposes would emphasize that life involves using energy from the surrounding environment to build up complex molecules, grow, and reproduce. Life always "feeds" off an external energy source—in the case of life on the surface of the Earth, that energy source is, of course, the Sun. A slightly more subtle definition would also emphasize that life is always associated with systems that are far from chemical equilibrium. For example, as a result of life processes the atmosphere of the Earth is rich in oxygen, a highly reactive gas. This does not represent chemical equilibrium. If there were no life on Earth, the oxygen would quickly become locked up in stable molecules such as carbon dioxide.* The fact that our neighbor Venus

---

*These subtleties, and others, are discussed in my *Deep Simplicity,* but we will not elaborate on them further here.

has a stable (equilibrium) atmosphere rich in carbon dioxide is compelling evidence that the planet does not harbor life.

Until recently, it was thought that essentially all the steps involved in the emergence of life took place on Earth, soon after our planet formed. But it has now become clear that at least the first step — the extraction of energy from the surrounding environment to build up complex molecules — took place, and is still taking place, in the clouds of dust and gas where stars form. Like the atoms which exist in the outer layers of a star, molecules in space can be identified by spectroscopy. But the key difference is that instead of being identified by the lines they produce in the spectrum of visible light, complex molecules, because they are bigger, are identified by their characteristic radiation at longer wavelengths, in the infrared and at radio wavelengths. The discovery of such molecules in space was delayed partly because the technology to search for them did not exist until the second half of the twentieth century, and partly because nobody expected them to be there, so nobody was looking.

The first molecules to be identified in space were discovered back in the 1930s, because they were easy to spot. But they hardly count as complex — a simple combination of carbon and hydrogen (CH) and the compound of carbon and nitrogen known as the cyanogen radical (CN). It wasn't until 1963 that another compound was identified — the hydroxyl radical (OH) — but the first really dramatic step forward came in 1968, when emission from ammonia, a four-atom molecule (NH_3), was detected coming from the direction of the center of our Galaxy. It was this discovery that started the ball rolling by encouraging astronomers to look for more complex molecules in space; such encouragement was necessary because in most cases they first had to decide what to look for, then measure the spectra of the appropriate molecules in the lab on Earth, before they could have a chance of identifying the radio spectra they detected from clouds of interstellar material. They soon discovered water (H_2O), and then the one which really set the bandwagon rolling, the organic molecule formaldehyde (H_2CO).

These discoveries came as something of a relief to biologists. Fossil evidence shows that life (single-celled life) already existed on Earth almost four billion years ago, less than a billion years after the Earth formed. A few hundred million years seems an uncomfortably short time for chemistry to progress from simple things like carbon dioxide and ammonia to things like proteins and DNA. But if complex organic molecules were around from the time our planet cooled, the speed with which life emerged is less surprising. In the past few years, astronomers have also identified these kinds of molecules in other galaxies, showing that their existence in interstellar space is a literally universal phenomenon, not one restricted to our Milky Way Galaxy.

As the name suggests, organic compounds are associated with life. All so-called organic molecules contain carbon atoms chemically bound to hydrogen atoms and in most cases to atoms of other elements as well. Originally, in the nineteenth century, it was thought that such compounds were exclusively associated with life, hence the name; but once it became clear that many organic molecules could be synthesized artificially in the proverbial test tube, the term *organic chemistry* became almost synonymous with *carbon chemistry*. But that does not mean that there is no link between organic chemistry and life; all living processes are associated with organic compounds, even though not all "organic" compounds are associated with life.

There are two reasons why carbon is so important to life. The first is that each carbon atom is capable of making four separate links (bonds) to other atoms, including other atoms of carbon, at the same time. Except for a couple of rather peculiar special cases, this is the maximum number of bonds that any atom can make, so carbon is able to make lots of connections to other atoms and can sit at the heart of a complex compound containing many atoms of different elements.* The other reason why carbon is so important is that it is relatively common. Apart from the hydrogen and helium which make up the bulk of the baryonic material in the Universe, the most common element is oxygen and the second most common is carbon, both produced by the process of stellar nucleosynthesis.

As a bonus, carbon atoms don't have to use up all four of their bonds by linking to four other atoms. They can also form double (or even triple) bonds, so that, for example, two carbon atoms can each use two of their bonds to link to one another by a double bond, leaving each of them with two bonds free to join on to other atoms. Carbon atoms can also form long chains, linked to one another like a spine, with other atoms and groups of atoms stuck on at the sides; and they can even make rings (most often with six carbon atoms "holding hands" in a loop) with other chemical stuff attached around the perimeter of the ring. So carbon is both common and eager to make many bonds with other atoms. With hindsight, it seems inevitable that there should be a lot of carbon compounds — organic compounds — in interstellar and circumstellar clouds, where the energy of starlight (including infrared and ultraviolet radiation) is available to drive interesting chemical reactions.

By 2005, more than 130 molecules had been detected in space, most of them in the giant molecular clouds where stars (and planets) are born. These range from the simple two-atom molecules such as nitric oxide (NO) and silicon monoxide (SiO), through three-atom

---

*There are other atoms which can make four chemical bonds each, most notably those of silicon. But there are eight times as many carbon atoms around as there are silicon atoms, and in any case the bonds made by silicon are weaker than those made by carbon.

varieties including hydrogen cyanide (HCN) and sulfur dioxide (SO_2), four-atom ammonia (NH_3) and acetylene (HC_2H), and five-atom formic acid (HCOOH, the active ingredient in bee stings and stinging nettles), to the larger organic molecules which are our primary interest here. Size isn't everything, and one of the largest molecules definitely identified in space to date consists of a rather boring chain of eleven carbon atoms with a single hydrogen atom on one end and a single nitrogen on the other; it is called cyanopentacetylene and has the chemical formula $HC_{11}N$. When it comes to life, complexity matters just as much as size, and it is much more exciting and significant to identify molecules which may be smaller than $HC_{11}N$ but which contain a richer variety of atoms arranged in more interesting ways. What we regard as interesting molecules, of course, are the ones which are used to make the building blocks of life; we can identify which ones those are by looking at the structure biochemists have discovered by taking biological molecules apart.

There are two kinds of large biological molecules which form the basis of life on Earth. One family, the proteins, provides the structure of your body (including things like hair and fingernails, as well as muscle), while other proteins, known as enzymes, directly control the chemistry of your body. The second family, nucleic acids (including the famous DNA, deoxyribonucleic acid) contains the coded instructions which tell the machinery of the cell how to make the different kinds of protein. The two kinds of molecules share one important common feature—they are both made of long chains in which molecular subunits are linked together by chemical bonds to provide a structure which contains a lot of information. And we do mean long—on a scale where the weight of a single carbon atom is twelve units, the molecular weights of proteins range from a few thousand units to a few million units.

In proteins (including enzymes), the subunits are molecules called amino acids. Amino acids themselves typically weigh in with little more than a hundred units on the same scale, which gives an indication of how many amino acids it takes to make a protein. But the importance of these molecules for life can be seen from the fact that half the mass of all the biological material on Earth is amino acids. The chemical unit that gives amino acids their name is built around a single carbon atom. One of the four bonds of the carbon atom is attached to a single atom of hydrogen, one to a set of three linked atoms known as an amine group (NH_2) and one to the carboxylic acid group ($COOH$) — hence the name *amino acid*. The fourth bond is free to link up to another carbon atom which itself has three further bonds with which it can attach to a variety of other groups.

Clearly, there is huge scope for an enormous variety of amino acids to exist, and very many have indeed been manufactured in the lab. But all the proteins found in living things on Earth are made up of different combinations of just twenty amino acids. The fact that all living things we know of use the same twenty building blocks in essentially the same way is a powerful piece of circumstantial evidence that all life on Earth stems from a single origin. We are all descended from some common ancestor, and although we cannot rule out the possibility that some completely different form of life also existed on Earth long ago, if it did it has left no traces, and no descendants. Proteins are without doubt life molecules, even though you could not say that an isolated protein molecule is "alive." You don't find proteins lying around in the world, produced by nonliving chemistry; you only find them associated with living things. But you do find amino acids lying around in the world, both the ones important for life and some of the ones that life makes no use of. In that sense, amino acids are nonlife. The trick that turns nonlife into life (whatever it is) seems to happen somewhere in the process that makes proteins out of amino acids. And that trick has to be associated with the complexity of proteins compared with amino acids — with the amount of information they contain.

This applies both to the long chains which make up the proteins in your hair, muscles, and other structural features of the body and to the proteins in which the chains are curled up into little balls, the so-called globular proteins which act as enzymes, encouraging some chemical reactions important to life and discouraging other chemical reactions that would be detrimental to life. It's particularly easy to understand the concept of information being stored in a protein as a code represented by the order of the amino acids along the protein chain, because the number of amino acids available for the biological machinery to use in this way, twenty, is so close to the number of letters in the English alphabet, twenty-six. We have no trouble grasping the idea that a lot of information (this book, for example) can be conveyed by a set of twenty-six letters (plus a few punctuation marks) presented in a well-chosen pattern that is really no more than a long chain, even if the chain has been cut up and arranged to make the lines on the pages. Similarly, proteins can be thought of as messages written in the twenty-letter amino acid alphabet. It is the information stored in this way in proteins that makes one protein chain suitable for forming part of a strand of hair, and another suitable for carrying oxygen round in your blood. But we do not wish to pursue the story of how biological molecules do their jobs here; what we are interested in now is where those biological molecules (and in particular, the first such molecules) came from. What we have learned is that if you have amino acids you are only one step away from life; so the next key question is, where do amino acids come from?

The twenty amino acids that are the stuff of proteins are themselves almost entirely made up of atoms of hydrogen, carbon, oxygen, and nitrogen (the four most common elements in the Universe apart from nonreactive helium) combined in different ways; just the odd atom of sulfur crops up in a couple of the amino acids. So in the 1920s the English biologist J. B. S. Haldane and the Soviet scientist A. I. Oparin independently suggested that when the Earth was young the energy available from the heat of the planet and from lightning flashes might have encouraged chemical reactions which led to the formation of amino acids from water and compounds such as methane and ammonia. Since the 1950s, these ideas have been tested in many experiments in which sealed vessels containing a variety of "atmospheres" have been subjected to electric discharges, ultraviolet radiation, and other sources of energy. The experiments generally do produce, if you wait long enough, a dark goo which does contain amino acids, including some of the ones that are the building blocks of proteins. But that doesn't prove that this is the way the first steps toward life took place on Earth.

Indeed, the wealth of complex molecules now identified in space suggests that the primordial Earth had a far richer brew of chemical components to start with. If laboratory chemists wish to synthesize amino acids, they don't start out with a flask containing water, methane, carbon dioxide, and ammonia and pass sparks through it for a few months. They start with things like formaldehyde, methanol, and formamide ($HCONH_2$) and do the job quickly and easily. All the reagents which are used in the laboratory to synthesize amino acids have now been found in giant molecular clouds (GMCs). At the very least, this means that the Earth could have been laced with such material (we shall discuss how shortly) soon after it was formed. There is also the tantalizing prospect that amino acids themselves may exist in GMCs. Indeed, there was a claim in 2003 that the simplest amino acid, glycine, had been detected in three GMCs. But in 2005 an exhaustive follow-up comparing new laboratory measurements of the radiation from glycine to the observations showed that the claim was mistaken. Nevertheless glycine, with a chemical formula of NH_2CH_2COOH, is such a relatively simple molecule that it will be no surprise if further searches do reveal its presence in some GMCs.

The latest development is that some of the significant organic molecules already known to exist in low concentrations in the clouds of gas and dust between the stars have now been identified in much greater concentrations in the dusty disks surrounding young stars. For example, a system known as IRS 46, which lies about 375 light years away from our Solar System, contains hydrogen cyanide at a concentration ten thousand times greater than that found in clouds of interstellar gas, and comparable concentrations of acetylene. The significance of this is that when hydrogen cyanide, acetylene, and water are mixed in the laboratory in a container where there are suitable surfaces on which molecules can grow, they produce a rich variety of organic compounds including both amino acids and one of the basic components of DNA, adenine. This material in the disk around IRS 46 is concentrated no more than 10 astronomical units out from the central star, equivalent to the region of our Solar System within the orbit of Saturn.

We may not yet have discovered the building blocks of proteins in space, but we have discovered the building blocks of the building blocks, a long way up the ladder from things like water, ammonia, and carbon dioxide. And the situation is at least as promising when we look for the building blocks of the other family of life molecules, the nucleic acids.

Like proteins, the nucleic acids are long chain molecules made up of many subunits strung together along a line, like beads on a necklace, with other chemical units stuck on at the sides. But the chemical subunits in the nucleic acids are simpler than amino acids, and there is less variety among them than there is among the twenty amino acids important for life. For a long time, this fooled biochemists into thinking that the nucleic acids are less important than proteins in the workings of the cell — perhaps merely a kind of scaffolding supporting the protein molecules. But they were wrong.

In chemical terms, both DNA and its close cousin RNA are made of sugar. The basic building block in both cases is a sugar molecule called ribose, which is made of four carbon atoms and one oxygen atom linked together in a pentagonal ring. Each of the four carbon atoms has two spare bonds which connect to other atoms or chemical groups. In both ribose and deoxyribose, on one side of the oxygen atom one of the carbons is attached to a hydrogen atom and to another carbon atom, which is itself attached to other atoms making the group CH_2OH; in ribose, all three of the other carbon atoms in the ring are attached to hydrogen and to an OH group. But in deoxyribose, instead of being attached to OH just one of the three is attached to H. There is one oxygen atom less in deoxyribose than there is in ribose, which is where the former gets its name.

These basic units are modified slightly in the nucleic acids. In both DNA and RNA, the final hydrogen atom in the CH_2OH group is replaced by a link to a chemical unit called the phosphate group, with a single phosphorus atom at its heart. The other side of the phosphate group is attached to another sugar ring in place of one of the hydrogen atoms in one of the OH groups. Each phosphate group provides a link between two sugar rings, so the spine of a nucleic acid is made up of alternating links in a chain that goes sugar-phosphate-sugar-phosphate-sugar-phosphate . . . On its own, that really would be boring and useless for anything much except scaffolding. But there's more.

In addition to its links with phosphate groups up and down the nucleic acid chain, each sugar ring is attached to one of five units known as bases, which stick out from the sides of the chain. There are, of course, many more than five chemical bases, just as there are many more than twenty amino acids; but only five are used in the nucleic acids. All five of these bases are built around hexagonal rings in which there are four carbon atoms and two nitrogen atoms. They attach to the sugar groups on the nucleic acid chain by replacing one of the OH groups attached to one of the carbon atoms in the sugar ring with a link to one of the nitrogen atoms in the base. The five bases are called uracil, thymine, cytosine, adenine, and guanine, and are usually referred to by the initial letters of their names. Only four of the five bases are found in each nucleic acid. DNA contains G, A, C, and T; RNA contains G, A, C, and U. Crucially, though, the bases can appear in any order along the molecules; the chemistry of a single strand of DNA doesn't "care" whether G sits next to A, C, or T. In each nucleic acid molecule, this means that instead of merely containing a boring repetition of the same sugar and phosphate groups over and over again, or an equally uninformative repetition of GACTGACTGACT . . . the nucleic acids contain information. They carry a "message" written in a four-letter alphabet along their spines. This is the message of the genes.

Any message can be written in a four-letter alphabet (or code) if the message is long enough. Indeed, any message can be written in a two-letter binary code, a string of ones and zeroes, like the one used by computers. The analogy we made before was between the message written in this book in the twenty-six characters of the alphabet and the information contained in proteins "written" in the twenty-letter amino acid alphabet; we might just as well have said that the message of our book is written in the two-letter binary code, since that is the code actually used inside the computer on which the book was written. If a string of os and 1s can convey all the information in a book and more, so can a string of Gs, As, Cs, and Ts or Us (the genetic code).

This is not the place to go into the fascinating details of how, inside the cell, the molecular machinery of life makes use of the information stored in the genetic code of DNA, translating it with the help of RNA to make the amino acids which are in turn assembled to make proteins,* but the fact that DNA does carry the genetic code is well enough known to make the relevant point here — if we are looking for molecules that could be the precursors to life, we should be looking not only for amino acids and the building blocks of amino acids but for ribose and the building blocks of ribose. Radio astronomers have indeed been looking for such building blocks — and in the early years of the twenty-first century they found them.

Specifically, they found features in the radio spectrum of the Sagittarius interstellar cloud, some 26,000 light years from Earth, corresponding to a sugar called glycoaldehyde (CH_2OHCHO). (These are exactly the same atoms, but arranged in a different way, as those in acetic acid and methyl formate, both also detected in interstellar clouds.) The sugar is detected both in warm regions of the cloud and in regions as cold as 8 K, and it is present in large quantities. It probably formed when shock waves spreading out from the sites where new stars were forming in the cloud provided the energy to allow the appropriate chemical reactions to take place. It's clear that the chemical reactions didn't stop there, since the observations also reveal the presence of ethylene glycol in the same cloud. Ethylene

---

*The story can be found in many books, including my own *In Search of the Double Helix* (1987).

glycol is a ten-atom molecule formed by adding two hydrogen atoms to glycoaldehyde, and it is much more interesting than cyanopentacetylene. It is one of the largest molecules yet identified in space and is most familiar in everyday life as the constituent of antifreeze.

The discovery of glycoaldehyde in particular is doubly significant. In general terms, it is another example of the fact that the kinds of molecules being discovered today in space are the same as the molecules that are produced in laboratory experiments specifically designed to synthesize prebiotic molecules; in particular, although glycoaldehyde is built around a core of just two carbon atoms, it is known to react readily with a three-carbon sugar to form ribose. All the evidence suggests that the chemistry going on in giant molecular clouds in our Galaxy is the same everywhere and is the chemistry that leads to the production of complex biomolecules including amino acids and nucleic acids. The two remaining questions are: How far can the buildup of complexity proceed in GMCs? and How do the complex molecules get down to the surface of a planet like Earth?

The intriguing answer to the first question is directly linked to the availability of carbon in interstellar clouds. One of the reasons why carbon is so common in the Universe is that carbon "burning" takes place only inside stars that have more than eight times as much mass as the Sun. About 95 percent of all stars have less mass than this, so nuclear burning in their interiors never gets further than the fusion of helium nuclei to make carbon nuclei. It's one thing to make carbon in the hearts of stars and another to get it first up to the surface of the star and then ejected into space; but the fact that stars achieve those tricks is shown by spectroscopic evidence for the existence of gaseous molecules and grains of dust in the expanding envelopes of material seen around many stars at a certain stage in their life cycle. This is the stage when the outer part of the star swells up and it becomes a red giant;

for historical reasons such a star is said to be on the "asymptotic giant branch"* and is sometimes referred to as an AGB star. Since the expanding clouds of material around AGB stars are moving so fast that they must disperse within a few thousand years, the evidence of molecules and dust in those clouds shows that complex structures are produced in them extremely rapidly by astronomical timescales.

Stars like the Sun (so-called Population I) actually begin their AGB phase of life with more oxygen than carbon. Computer simulations show that carbon made in the core of the star is dredged up to the surface by convection, where it builds up in the thin outer layers of the star until the number of carbon atoms exceeds the number of oxygen atoms (if the process goes on long enough; not all stars get to this point). It is one of the curiosities that help make carbon so important for life that carbon atoms can link up with other carbon atoms in a variety of ways if the conditions are right, so although most of the carbon combines with oxygen to form carbon monoxide (CO), and some links up with nitrogen to make CN, there is some left over to form such varieties as C_2 and C_3. Stars that show spectral features corresponding to these substances are called carbon stars, although they are not, of course, entirely made of carbon.

A typical AGB star swells up to a diameter hundreds of times the diameter of the Sun and reaches a brightness several thousand times that of the Sun during this phase of its life. At such a size the pull of gravity at the surface of the star is very weak, and the outward pressure from the star's radiation is very strong. As a result, material escapes from the surface of the star in a stellar wind, carrying away the equivalent of as much as one-ten-thousandth of the mass of the Sun in a

---

*This "branch" is a region on a standard plot of stellar brightness against star color, called the Hertzprung-Russell diagram.

year. That doesn't sound like a lot, but over a thousand years it means that the star loses one-tenth of a solar mass of material, equivalent to 33,000 times the mass of the Earth. Because of the low temperatures in this expanding cloud of stuff, many stable molecules can form. More than sixty different kinds of molecule have been seen in the spectra of AGB stars, including simple organic compounds such as H_2CO and CH_3CN, ring molecules including the triangular pro-pynylidine (C_3H_2), and our old (but boring) friend $HC_{11}N$.

The solid particles definitely identified in the material that surrounds AGB stars include silicates and silicon carbide (SiC). These solid particles absorb starlight and reradiate the energy they have absorbed in the infrared part of the spectrum; there is so much dust around some AGB stars that they cannot be seen at all using optical light and are only identified with the aid of infrared telescopes. Because infrared radiation is absorbed by the atmosphere of the Earth, as we mentioned earlier, such stars can only be identified from infrared satellite observatories or from telescopes on high mountaintops above the obscuring layers of the atmosphere. As a result, the investigation of the kinds of molecules and (in particular) solid particles in the circumstellar envelopes of AGB stars is a new branch of astronomy in which the evidence is still open to a variety of interpretations, and there is no clear single explanation for the observations. The material in the circumstellar clouds has to be investigated by comparing the infrared spectra of these stars with the spectra of minerals studied in the lab; but there is always the possibility that the stellar environment may have produced substances which are not known here on Earth. There is, though, still a lot we can infer about what goes on in these clouds. Although some of the conclusions we will present here are slightly speculative, part of what we *think* we know, they offer tantalizing clues to the origin of life.

Not all giant stars have atmospheres dominated by carbon. In some cases, the number of carbon atoms never exceeds the number of oxygen atoms. In either case, the junior component gets locked up as CO, although it may eventually be involved in other reactions. In the oxygen-rich stars, the compounds produced are mostly oxides, and in carbon-rich stars organic compounds are produced. Both kinds of material, however, are dispersed through space and mix with primordial hydrogen and helium to form the raw material from which the next generation of stars and planets forms.

The most important oxides (apart from water) are the silicates, oxides of silicon sometimes combined with other elements. Ordinary sand is mostly made up of grains of the simplest silicate, silicon dioxide (SiO_2); silicates are the most common minerals in the Earth's crust and have been detected in the spectra of more than four thousand AGB stars, so there is no mystery about their origin. Other oxides associated with AGB stars, identified in the era of orbiting infrared observatories, include corundum* and spinel, a mixture of oxides of aluminum combined with magnesium and iron. Intriguing though these discoveries are, however, what we are really interested in here are the organic compounds associated with carbon-rich stars.

Even in carbon-rich AGB stars, the most common solid firmly identified in the dust grains is silicon carbide, SiC, which has been found in 700 carbon stars. But the spectral features attributed to SiC are weaker in carbon stars that are further along in their life cycle, which tells us that the silicon carbide is no longer a dominant component of the dust. This is where the element of speculation comes in.

---

*An oxide of aluminum which is the second-hardest naturally occurring mineral after diamond; it is more commonly known on Earth in the form of rubies and sapphires, though there aren't rubies and sapphires in the atmospheres of AGB stars — the material is in a softer state known as the amorphous phase.

There are strong but as yet unidentified emission features in the spectra of carbon stars that have passed beyond the AGB stage. Just twelve stars showing the first of these features had been found by 2004, and there is no obvious explanation for this spectral feature except that it is being produced by some form of carbon. But the feature is spread over a broad range of infrared wavelengths, and it does not contain the sharp spectral lines that would be associated with individual varieties of molecules such as SiC. A second feature found in the spectra of carbon stars at a different wavelength (or rather, spread over a different range of wavelengths) shows many similar properties to the first one. Both features may be explained by the combined effect of infrared radiation from very many carbon ring molecules linked together, although there is as yet no proof that this is what is going on.

Compounds containing these carbon rings are known as aromatic compounds, because they often have a pronounced scent — although it is not always a pleasant fragrance. The archetypal example is benzene, whose molecular structure (C_6H_6) consists of six carbon atoms linked in a hexagon with one hydrogen atom attached to each carbon atom; the structure is known as a benzene ring and lies at the heart of all the molecules that chemists call aromatics, sometimes with an atom of a different element substituted for one of the carbon atoms in the ring.

In one example of this kind of molecule, the odd carbon atom is replaced by an oxygen atom, making a pyran ring (C_5O). Pyran rings easily form long chains in which each ring is attached to its neighbor at either end by an oxygen atom that acts as a bridge between the two rings; such long chains are called polymers, in general, or polysaccharides in this particular case. Once a few such chains exist, they will tend to grow by latching on to more carbon and oxygen atoms, if they are available, and turning them into more pyran rings. And if one of the chains snaps, you then have two polysaccharide chains. The ability to grow and reproduce is a key property of life, and although polysaccharides are not alive, their example shows how this key property could have emerged naturally as the chemistry got more complicated.

Crucially, acetylene (C_2H_2), the basic building block for benzene and other aromatic compounds, is one of the molecules already identified in these clouds. The broad features in the spectra of carbon stars are in the right part of the infrared spectrum to be associated with stretching and bending of the C-H and C-C bonds in benzene rings, which produce features collectively referred to as aromatic infrared bands, or AIBs. But the features in the spectra from space could be explained only if very many benzene rings were linked together in sheets and chains of hexagons, a mass of carbonaceous material containing at a minimum hundreds of carbon atoms. Such combinations of a large number of benzene rings are known as polycyclic aromatic hydrocarbons, or PAH. It is very difficult to simulate the conditions in which these complex structures exist in deep space in order to measure their spectra in the laboratory. Nevertheless, by probing synthetic PAH structures with a laser beam in a near-vacuum at temperatures close to absolute zero, experiments carried out in 2002 at the University of Nijmegen in the Netherlands provided the best evidence yet that the identification is correct. PAH can also be linked to carbon-based chains of smaller molecules, and those chains can act as bridges to other sheets of PAH. There is a very common substance found on Earth which has essentially this structure — coal.

Strong circumstantial evidence that this is the right explanation of the broad infrared emission bands seen in carbon stars comes from within our own Solar System. Meteorites are leftover fragments from the formation of the Solar System that sometimes fall to Earth; they contain material that is representative of the solid material in the cloud of gas and dust from which the Solar System formed. The most common organic material found in meteorites is kerogen, a coal-like material that is the solid organic component of oil shales and produces hydrocarbons similar to petroleum when heated. This does not mean that coal and oil come from space. We are suggesting that PAH may be one of the ingredients from which life developed, while coal and petroleum are the remains of once-living things, so they are at opposite ends of the story of life — a case of "coal to coal," rather than "ashes to ashes."

When NASA's Deep Impact space probe was deliberately crashed into a comet, Tempel-1, in 2005, it stirred up cometary material which was analyzed at infrared wavelengths using the Spitzer telescope back on Earth. To the surprise of even many astronomers (but not those who had been following the story outlined here), the spectra from this cometary material revealed the presence of silicates, carbonates, claylike materials, iron-bearing compounds, and aromatic hydrocarbon compounds similar to those found in barbecue pits and exhaust emissions. To those who had been following the story, this was a neat and satisfying discovery which slotted another piece of the jigsaw perfectly into place in the emerging picture of how the materials essential for life came down to Earth.

Meteorites also have something else to tell us about the origin of life. We have mentioned that life makes extensive use of carbon, hydrogen, oxygen, and nitrogen, the four most common reactive elements there are. Other elements are present in much smaller quantities in the molecules of life, as you would expect from their relative scarcity; but there is one curious exception. Phosphorous, as we have seen, is a key component of the nucleic acids, and it also turns up in other molecules of life in what may initially seem surprising amounts. To put this in perspective, in the Universe at large there is one phosphorous atom for every fourteen hundred oxygen atoms; but in bacteria (single-celled organisms that are in many ways the basic unit of life) there is one phosphorous atom for every seventy-two oxygen atoms, making it the fifth-most-important biological element in terms of mass.

The reason is that phosphorous has an unusual facility for forming links with other atoms. Through a quirk of quantum mechanics affecting the behavior of the electrons in the phosphorous atom, it is sometimes possible for a single atom of phosphorous to make bonds with five other atoms at the same time. This enables it to form a component of a huge number of molecules, and to link other chemical units together in complex and interesting ways. Once you know this, it is no surprise to find that phosphorous is a key component of the complexity of life; its ability to form many chemical bonds in this way more than makes up for its scarcity (and contrasts it neatly with helium, which makes up 25 percent of the baryonic material in the Universe but doesn't form stable bonds with anything, so is completely absent from the molecules of life). Any gardener or farmer can tell you how important phosphate fertilizers are to plants.

What has this got to do with meteorites? Many meteorites contain phosphorous, typically bound up in a mineral form with iron and nickel. In 2004 researchers at the University of Arizona carried out a simple experiment in which one of these minerals, known as schreibersite, was put in contact with ordinary water at room temperature. The resulting chemical reactions produced a variety of phosphorous compounds, including the oxide P_2O_7, which is used in several biochemical processes and is similar to a component of a compound called adenosine triphosphate (ATP) which is used to store energy in all living cells. One of the roles of ATP is to provide the driving force for muscle contraction. Yet again, we have found evidence for the presence of the building blocks of the building blocks of life in space. Meteorites have also been found to contain amino acids (confirming that these building blocks of proteins were already present in the material from which the Solar System formed), carboxylic acid, and sugars, including glycerin, a sugar used by all cells on Earth today in building cell walls, and glucose, a hexagonal ring molecule ($C_6H_{12}O_6$) important in respiration.

There's something else about the molecules found in meteorites which provides a link between life on Earth today and the origins of life molecules in space. Complicated molecules like amino acids and the more interesting sugars have distinctive, three-dimensional shapes and can usually exist in either of two forms which are mirror images, like the left and right hands of a pair of gloves. These are usually referred to as the left-handed and right-handed isomers of the molecule. The "handedness" is defined by the way the molecules affect polarized light. Polarized light can be thought of as like a series of vertical ripples running along a stretched rope; the effect of the left-handed and right-handed isomers is to change the angle the ripples make, so that what were upright ripples lean a little to the left or to the right. When chemists synthesize such molecules from the basic atoms of which they are composed, they get equal quantities of the left-handed and right-handed forms. The laws of quantum chemistry do not favor either variety over the other. But life on Earth uses almost exclusively left-handed amino acids to build proteins and right-handed sugars to build nucleic acids. A DNA molecule, for example, can no more make use of left-handed deoxyribose than you can fit your left glove on your right hand.*

---

*This is the secret of some low calorie sweeteners. If they are made from left-handed sugars, the body cannot use them, even though they still taste sweet.

The first thing this tells us (confirming other evidence) is that all life on Earth today is descended from a common ancestor. If the ancestral life form — the original cell, perhaps — just happened to use these isomers, all its descendants would continue to do so, regardless of the presence of the mirror-image versions of these molecules in the environment. So until recently there seemed to be the intriguing possibility that if life-forms similar to us were found on nearby planets, they might use right-handed amino acids and left-handed sugars, or have both kinds of molecule with the same handedness. This produced several science-fictional variations on the theme, with stories of stranded space travelers starving amid plenty because their metabolisms could not cope with the food found on other worlds. But in the late 1990s, astrobiologists discovered that the amino acids found in meteorites are also left-handed. The asymmetry already existed in the precursors of the molecules of life before the Solar System formed.

There are two ways to provide a surplus of one handedness of molecules over the other. Either you make more of one isomer in the first place, or you destroy the second isomer after it has been made. In laboratory experiments, molecules of one handedness are eliminated preferentially by the effects of circularly polarized light. (It's hard to picture this, but circularly polarized light behaves as if the light waves were rotating as they move through space, like a screw rotating as it is being driven into a piece of wood.) The last piece of this particular puzzle fell into place when astronomers using the Anglo-Australian Telescope at Siding Spring Mountain in Australia detected circularly polarized light (in the infrared) coming from the Orion molecular cloud. This is a region where stars are forming and organic molecules have been detected. It seems certain that the circularly polarized light in the region will imprint a preferred handedness on these organic molecules, before parts of the cloud collapse to form new stars and planets.

The implication is that a characteristic pattern of handedness will be imprinted on all the material from which a group of stars forms together. But since circularly polarized light can itself be either left-handed or right-handed, depending on how it rotates, molecules in different interstellar clouds (or even in different parts of the same GMC) may be affected in different ways. So although it is now certain that if there are amino acids anywhere else in the Solar System they will be left-handed, like those on Earth, different variations on the handedness theme could still exist in different parts of the Galaxy. Those science-fiction stories could turn out to be right after all, if the space travelers venture far enough.

This is enough evidence for now that the building blocks of life already existed in the cloud of material from which the Sun and its family of planets, including the Earth, formed. We have also seen that this material can get down to the surface of the Earth inside meteorites. That could have been enough on its own to kick start life on Earth. But there is an even better way to get the organic material down to Earth, which would have been even more effective when the Solar System was young — inside comets.

Comets are no longer the mysterious objects that aroused superstitious fear in our ancestors. The process of taming them began in the eighteenth century, when Edmond Halley correctly predicted the return of the comet that now bears his name, showing that they are ordinary members of the Solar System under the influence of gravity and subject to the same laws as the planets and other objects moving around the Sun. In recent times, as well as being studied spectroscopically from the ground, comets have been visited by flyby space probes, and in the dramatic case of Tempel-1, in 2005 a probe was deliberately rammed into the comet, providing close-up pictures and creating an outburst of material which (to no one's surprise) revealed the presence of large amounts of water in the comet. Comets are, indeed, best regarded as dirty space icebergs — they used to be called dirty snowballs, but we now know that they are much more solid than that name suggests. A lot of the dirt in a comet is organic material, carbon compounds, and although this dirt is only a small proportion of the mass of a comet, comets are so big (Halley's Comet itself has a mass of about 300 billion tonnes) that even a small proportion of the total adds up to a lot of material by human standards. And that's just for a single comet.

There are two families of comet, distinguished only by their orbits. So-called short-period comets travel in elliptical orbits which stretch across roughly the same region of the Solar System as the orbits of the planets. Halley's Comet, for example, goes out a little beyond Neptune and comes in toward the Sun a little closer than Venus, taking about seventy-six years to complete each orbit. For most of that time, it is nothing more than, literally, a dirty great lump of ice. But during its close passages past the Sun, it gets hot enough for material to evaporate from its surface and stream out in a long tail, the visible defining feature of a comet that so impressed our ancestors. The next return of this particular comet, in 2061, is likely to be unremarkable, because it will be in the wrong part of the Solar System to provide a good show from Earth; but in 2134 Halley's Comet will pass within 14 million kilometers of our planet, and will provide a spectacular show in the sky. There are now more than a hundred known short-period comets whose orbits have been determined accurately.

Long-period comets have much more elongated orbits, taking them much farther out from the Sun. The arbitrary distinction between long- and short-period comets was set historically at an orbital period of two hundred years, but this historical accident is a little misleading. Although accurate orbits have been determined for more than five hundred long-period comets, the really important distinction is between comets that have any known orbital period and those which have such long, thin orbits that it is impossible to calculate a precise orbital period at all. They appear from the depths of space, whip once around the Sun, and disappear again, as far as anyone can tell, forever.

Orbital calculations show that all the comets with known orbital periods can be explained as former ultra-long-period comets that have been captured and trapped in the region of the planets by the gravitational influence of Jupiter. But the calculations also show that only one in a million of these "wild" comets gets captured in this way. For every short-period comet, like Halley's Comet, that we see, there must have been a million visitors from the outer regions of the Solar System. So there must be a vast reservoir of comets somewhere out in space providing a steady supply of visitors to our part of the Solar System, in order to account for the hundreds of comets with known orbital periods.

The calculations show that these wild comets all have orbits which started out about 100,000 astronomical units from the Sun, literally halfway to the nearest star. (For comparison, the farthest Neptune gets from the Sun is just over 30 AU.) This gave rise to the idea that the Sun is surrounded by a vast cloud of comets at this distance, a cometary reservoir which became known as the Öpik-Oort Cloud, after the two astronomers who pioneered the idea. For half a century, there was scope for debate about whether this cloud really existed, and if it did exist, how the comets got there in the first place. But since the 1990s, studies of dusty disks around young stars such as Beta Pictoris have settled the debate and shown, crucially, that it really was "in the first place," at the origin of the Solar System, that the cometary reservoir got filled up.

Computer simulations of the formation of the Solar System which include an allowance for a disk like that around Beta Pictoris, containing more mass than the Sun itself, show that although most of the material in the disk is lost into space as the system settles down, several hundred Earth-masses of stuff, more than the mass of all the planets put together, remains in the form of comets. Some of these are in an extended disk around the Sun beyond the orbit of Neptune, and some, a minimum of a hundred Earth-masses of material, in the Öpik-Oort Cloud. That's enough material to make 2,000 billion objects the size of Halley's Comet — a virtually inexhaustible reservoir as far as the Sun is concerned. If even twenty of these comets fell in toward the Sun each year — a rate far in excess of the rate today — then in the lifetime of the Sun to date only 5 percent of the original number would have been lost from the reservoir.

A typical comet in the cloud is orbiting the Sun at a leisurely 100 meters per second, only ten times faster than an Olympic sprinter, and does so for billions of years. Just occasionally, some disturbance in the cloud (perhaps the gravitational influence of a nearby star, or some interaction between two comets) will send one or more of them first drifting, then hurtling, into the inner Solar System, where the vast majority whip around the Sun, like cars negotiating the high-speed banking at the Indy 500, before disappearing back out into space. Just one in a million of these visitors gets deflected by Jupiter into a short-period orbit around the Sun; sometimes the comet will collide with Jupiter (as in the case of Shoemaker-Levy 9 in 1994) or one of the other planets. Depending on the size of the comet, a collision with Earth could be a local disaster (as seems to have happened in the Tunguska region of Siberia in 1908) or a global catastrophe (as seems to have happened at the end of the Cretaceous Period, 65 million years ago, the time of "the death of the dinosaurs"). But as well as bringing death to Earth, comets may also have brought life.

It is clear that comets are composed of pristine material typical of the cloud of gas and dust from which the Solar System formed. This means that they are likely to contain all the ingredients seen in GMCs, plus more complex molecules that we have not yet identified in space. Since some meteorites contain amino acids, for example, it would be astonishing if at least some comets do not also contain amino acids. It is also clear that there must have been many more comets wandering through the inner Solar System when it was young, just after the planets formed and before everything settled down into a stable state, than there are today. The simulations we discussed earlier show how in this early phase of the history of the Solar System many comets were ejected out into space while a comparable number fell in toward the Sun, where they were swept up by the planets. As we have seen, the battered face of the Moon bears mute testimony to the number of impacts that occurred a little more than four billion years ago, and whatever was battering our neighbor the Moon must also have battered the Earth. It's very likely that a substantial proportion (perhaps all) of the Earth's water was deposited on the planet during this bombardment, along with other cometary material. It's possible that organic material could have survived such impacts, but there is no need to speculate about this since such material would also have had a far gentler passage down to Earth.

The most characteristic feature of comets is that they lose material when they pass close by the Sun. A comet can also break apart into smaller pieces, ejecting even more material in the process, if the heat from the Sun makes the icy materials inside it turn to gas, cracking the comet and breaking it apart. By the time it has orbited the Sun a few times, a short-period comet has left a trail of dust along its entire orbit; when the Earth passes through such a dust trail we see showers of meteors, streaks of light in the sky caused by the dust grains burning up in the atmosphere. Two such displays of "shooting stars" occur in November (the Leonids) and August (the Perseids) each year. But some of the grains do not burn up. The smaller ones settle gently down through the atmosphere and reach the surface of the Earth intact.

Even today, cometary dust grains gathered up in this way contribute about 300 tonnes of fresh organic material to the Earth each year, while meteorites contribute about 10 kilos of organic stuff safely sealed inside the rocky objects as they dive through the atmosphere. Just after the primordial bombardment of our planet ended, there must have been a lot more of this dust around in the inner Solar System. A conservative estimate, based on the simulations we have already described, is that about 10,000 tonnes of organic material, brought to the inner Solar System by comets but essentially the unadulterated stuff of GMCs, reached the surface of the Earth each year at that time. In a hundred million years or so, about the time it took for life to make its first marks on Earth, that amounts to an awful lot of organic material — polyatomic molecules containing carbon. Given those starting conditions, it is hard to see how life could have failed to get a grip on the surface of the Earth; the remaining question is, just how close to being alive was the cometary material before it even reached the Earth?

The basic unit of life is the cell. The chemistry of life involves molecules such as proteins and nucleic acids, assembled from subunits like amino acids and sugars; but that chemistry only takes place inside the protective wall of a cell, shielded from the external environment. If odd molecules of DNA and proteins were floating about loose in the sea, there would be very little chance of them ever getting together to do the things that make life interesting. Clearly, the emergence of life depended on the key molecules being confined somehow in a place where they could interact with one another. There have been many suggestions as to the site where life might have started — one intriguing possibility is that the key molecules were trapped in layers of claylike material. The one we are about to outline is not unique, and not proven; but it is a front-runner, and it fits all the known facts.

Most cells are tiny, perhaps a tenth or a hundredth of a millimeter in diameter. Your body contains about a hundred thousand billion of these cells — several hundred times the number of bright stars in the Milky Way — working together to make you what you are; but single-celled organisms such as bacteria can manage quite well on their own. The most important feature of the cell is the membrane which surrounds the watery fluid, known as cytoplasm, in which the chemistry of life takes place. This barrier against the outside world is only a few ten-millionths of a millimeter thick, and it allows certain molecules to get in (essentially, "food") and certain molecules to get out (waste products). The membrane selects these molecules on the basis of their size and shape. In order to allow different molecules to pass through in different directions it is not a uniform structure like a brick wall with holes, but has a definite "inside" and "outside." The simplest kinds of single-celled organisms have cells with the least internal structure, and it is reasonable to assume that these represent the primordial form, the original single-celled species.

Fossil evidence shows that complex multicelled organisms did not appear on Earth until about 600 million years ago, three *billion* years after single-celled life appeared. Creatures like ourselves evolved even later. But here we are only interested in the *origin* of life, so these primitive single cells are the ones to concentrate on. They are always surrounded by a chemically complex membrane which is made of long chains (polymers) of molecules known as amino-sugars. The chains are joined to one another by other chemical units (short chains of amino acids called peptides) to form a net vaguely reminiscent of a string bag. The point we want to emphasize is obvious from the name of the compounds involved — the subunits from which this cell wall is made are amino acids and sugars, both likely components of the cloud from which the Solar System formed, and both likely components of comet dust.

In the late 1990s and at the beginning of the twenty-first century, NASA scientists and researchers at the University of California, Santa Cruz, carried out experiments which suggested that membranelike structures could form in association with the icy dust grains that exist in GMCs. They made ice grains by freezing a gaseous mixture of some of the simpler compounds known to exist in GMCs — water, methane, ammonia, and carbon monoxide plus the simplest alcohol, methanol — onto little squares of aluminum cooled to -263 °C, similar to the way a layer of frost forms on a car windscreen on a chilly winter night. The ice grains were then irradiated with ultraviolet radiation, mimicking the radiation from hot young stars. The end products included more complex alcohols, aldehydes, and a large organic compound known as hexamethylenetetramine, or HMT. But the really dramatic discovery came when the team allowed the material to warm up in liquid water. They found that some of the constituents spontaneously formed themselves into little hollow spheres, dubbed vesicles, between 10 and 40 millionths of a meter across — about the same size as red blood cells.

The explanation is simple, once you understand the nature of some of the complex organic molecules produced by the effect of ultraviolet light on the ices. These particular molecules are called amphiphiles, and they behave in a similar way to the active molecules in laundry detergent. The molecules have a distinctive "head" and "tail" structure, with the tail being repelled by water and the head attracted by water. The effect in detergents is that the tails get buried in grains of dirt, so that the detergent molecules surround the dirt grains and (with the aid of a little shaking) float them loose from the material being washed. With nothing available to bury their tails in, however, the amphiphiles in these simulated space conditions form a double layer, with tails on the inside and heads on the outside. These layers then naturally curl up to make little hollow spheres. As a bonus, the layers absorb ultraviolet light, so that the interiors of the vesicles are little havens where chemistry can proceed without interference from outside.

The conservative interpretation of these new discoveries is that vesicles from comet dust would have been floating around in the waters of the young Earth, laced with other organic material from space, and that in some warm little pond, things like amino acids and sugars trapped inside them began the processes that would lead to life. The more radical interpretation, which I favor, is that inside the icy bulk of a comet, warmed by the radioactive decay of short-lived isotopes produced by a supernova explosion, vesicles formed in little puddles and became filled with complex organic molecules, up to and including the molecules of life. Apart from anything else, this extends the time available for chemistry to take the step from nonliving to living from the couple of hundred million years available on the surface of the Earth to several thousand million years. Even if the comet later froze solid, the vesicles would be waiting, ready to thaw out and deliver the seeds of life to a planet when they became part of the rain of cometary dust in the inner Solar System, and in similar systems.

These ideas are still seen as daring in the first decade of the twenty-first century. But it's worth noting that the first person to suggest such a scenario was the astrophysicist Fred Hoyle, in the 1970s. At that time, Hoyle's ideas were essentially laughed out of court, not least because, together with his colleague Chandra Wickramasinghe, Hoyle went so far as to suggest that diseases such as influenza might be being brought to Earth by comet dust today. It now looks, however, as if Hoyle's hypothesis was more right than wrong, even if he got a little carried away.* The idea that cells came first, followed by enzymes and only then by genes, has an even longer pedigree, and goes back to the work of A. I. Oparin in the 1920s, although the suggestion that the first cells emerged in space came much later. It has been said that the fate of a new idea in science is first to be dismissed as ridiculous, then to become a revolutionary new theory, and finally to be regarded as self-evident;† we may be at the second of these stages with the idea that life itself originated in deep space and was brought to Earth by comets.

This is the most extreme possibility being considered seriously today. My personal answer to the question "Where did life originate?" would be, "In the icy material of GMCs, in the material from which stars and planets then formed." But it is a sign of the times that the *conservative* answer today would be, "In some warm little pond on Earth, where complex organic molecules brought to Earth by comets took the crucial step to reproduction." Either way, there is no doubt that all Earth-like planets will be seeded with the same kind of organic material when they are young. This means that life is likely to be common across the Universe, and that all life is based on the same basic

---

*And if I am permitted a small self-congratulatory pat on the back, I said as much in my book *Genesis,* published in 1981.

†Or as the German philosopher Arthur Schopenhauer put it: "All truth passes through three stages. First, it is ridiculed. Second, it is violently opposed. Third, it is accepted as being self-evident."

components, amino acids and sugars, although life elsewhere may use different amino acids and sugars from life on Earth. The possibility of *intelligent* life elsewhere is, though, another question, and one beyond the scope of the present book. The last big question we want to address here is indeed the last question — how will it all end?

10

How Will It All End?

Judging from the evidence of life on Earth — the only evidence we have — once life gets a grip on a planet it is pretty resilient. The prospects for human civilization are clouded by many uncertainties of our own making, including wars, anthropogenic climate change, and degradation of the environment;* whether we survive these hazards is not a topic for scientific debate but a matter of political will. There is compelling scientific evidence, for example, that human activities are making the world warmer at an uncomfortable rate, but whether we do anything about this is a political decision; equally, we have the scientific and technological knowledge to feed adequately an even greater human population than the Earth already carries, but people continue to starve in large numbers because of political decisions. Whatever the outcome of those decisions, though, and whatever happens to human civilization over the next few hundred years (or the

*A good place to get a feel for these hazards is in *Our Final Century* (2003), by Martin Rees.

246

next few hundred thousand years), life will go on. After all, the oldest forms of life on Earth, single-celled bacteria, have been around for nearly four billion years, surviving everything the environment has thrown at them.

The biggest natural threat to our kind of life on Earth may be, ironically, the same process that brought life to Earth in the first place — impacts from space. The geological record shows that there have been many occasions when large numbers of species (not just individuals, but whole *species*) on our planet have been wiped out in so-called extinctions, some of which must have been associated with meteorites or comets striking the Earth. The most famous of these mass extinction events, the "death of the dinosaurs," occurred 65 million years ago, and meteorite impact is thought to have played a part (perhaps the dominant part) in it. This was the latest of the five biggest mass extinctions that have occurred since the earliest fish evolved. The first happened about 440 million years ago, the second just over 360 million years ago, the third (and largest) about 250 million years ago, and the fourth some 215 million years ago. Although it wasn't the biggest of these extinctions, the event that occurred 65 million years ago (known as the Cretaceous-Tertiary event, from the names of the two geological intervals separated by the disaster) is the one we know most about, since it was the most recent. More than 70 percent of all species alive on Earth were wiped out at the end of the Cretaceous period, and a similar catastrophe today would almost certainly bring an end to the human species, along with many others. Some of the other big five events were even more destructive. But the point we want to make here is not that so many species have been wiped out so often during the history of life on Earth, but that in spite of this, life goes on. New species evolve and adapt to the changed circumstances after every disaster, and have done so for roughly four billion years. So what would it take to bring an end to life on Earth?

The only certainty seems to be that our planet will become uninhabitable (even for bacteria) when the Sun swells up to become a red giant as it nears the end of its life. This is a well-understood process, and takes us firmly back to what we think we *know*, rather than what we *think* we know.

Prosaically, red giant stars get their name because they are red and they are large. All stars like the Sun suffer this fate as they run out of nuclear fuel. As long as there is enough hydrogen in the core of the Sun to provide energy to hold the outer layers of the star up against the pull of gravity by converting protons (hydrogen nuclei) into helium nuclei in the way we described earlier, all is reasonably well with the Sun, and conditions are reasonably favorable for life on Earth. (Only "reasonably" because during this long phase of its life the Sun will actually get slightly hotter, and this may make things uncomfortable for our kind of life in the not-too-distant future, perhaps within a few hundred million years, even if bacteria survive.) In round terms, the Sun started out with enough fuel for this process to continue for about ten billion years, and we are at present rather less than halfway through this period of long-term stability. That's the good news.

When the hydrogen fuel in the core of a star like the Sun is exhausted, it can no longer hold itself up against the pull of gravity, so it shrinks. But as the core shrinks this releases gravitational energy which makes the core hotter, and the extra heat from below makes the outer layers of the star expand into space. Because the core of the star has gotten hotter, more heat escapes from its surface. But because the star has expanded, the area of its surface has also increased, and the net effect is that even though the total amount of heat crossing the whole surface of the star has gone up, the amount of heat crossing each square meter of the surface has gone down. So the temperature of the surface goes down, even though more energy is escaping into space. That is why red giants are red, not yellow or blue. But this first experience of gigantism doesn't last forever. The extra heat in the heart of the star ignites helium burning, in which helium nuclei are fused together to make carbon nuclei. The energy released allows the core to expand slightly and cool off a little, while the outer layers shrink back from their extended state.

When all the helium in the core is used up (which takes only about a hundred million years, nowhere near as long as the hydrogen-burning phase of the star's life), the same sort of thing happens again. The core shrinks and gets hotter while the outer layers expand even more, and the star becomes a supergiant. Carbon is the end of the line for nucleosynthesis in stars like the Sun, and the processes which build up the heavier elements take place, as we have seen, only in much more massive stars. For a time, though, a star with about the same mass as our Sun can maintain itself as a supergiant with an inert carbon core by burning hydrogen into helium in a shell around the core. This steadily makes the core more massive and compact, but the outer layers of the star continue to expand, losing material into space. Eventually, when all this fuel is exhausted, the star will cool and con-tract into a white dwarf, a stellar cinder with nearly as much mass as our Sun packed into a volume no bigger than that of the Earth.

Many popular accounts of this process (and even some textbooks, whose authors ought to know better) tell us that the fate of the Earth is to be swallowed up by the Sun when it becomes a red supergiant in about 7.5 billion years from now. The end of life on Earth in these scenarios is estimated as occurring in about 5.5 billion years, when the Sun becomes twice as bright as it is today and scorches our planet. But the mistake the authors of these scenarios make is that at each step in their calculations they use the mass of the Sun today, and when they make comparisons with observations they look at stars, including red giants, which have the same mass now as the Sun has. They take no account of the way the Sun will lose mass as it ages, particularly when it begins to expand. A red giant which has one solar mass now had to have started out with substantially more mass, and a star which starts out with one solar mass will end up with much less. Even a rough calculation shows that the Earth will never actually be engulfed by the Sun, although it cannot survive as a home for life if nature takes its course; more knowledgeable prognostications have long made this clear. But now we can do even better. A more accurate forecast of the fate of the Sun and Earth has been carried out by some of my colleagues at the University of Sussex, and this provides the best guide yet to the long-term fate of our planet (Schröder, Smith, and Apps, "Solar Evolution and the Distant Future of the Earth").

At present, the Earth orbits at a distance of about 150 million kilometers from the Sun (strictly speaking, 149.6 million kilometers from the *center* of the Sun; the Sun's radius is 1.4 million kilometers, so we are "only" 148.2 million kilometers from the surface of the Sun). The calculations (and comparison with giant stars observed in the Galaxy today) show that when the Sun first becomes a red giant, even after allowing for mass loss it will expand to a radius of 168 million kilometers, which might seem sufficient to engulf the Earth. But by that time it will have lost so much mass that its gravitational grip on the planets will have loosened significantly, and the Earth will have drifted out to an orbit with a radius of 185 million kilometers. Because so much mass is lost from the outer part of the Sun (as much as 20 percent of the star's initial mass by the time it becomes a red giant), there will be a lot less fuel to drive hydrogen burning in the later stage of expansion, and in fact the Sun itself will never become a *super*giant — in the second phase of expansion its radius will increase only to 172 million kilometers, not much bigger than in the first red giant phase, and still not sufficient to engulf the Earth. By then, the total mass loss will be about 30 percent of its starting mass, and the Earth's orbit will have expanded out to a radius of 220 million kilometers, almost 50 percent greater than it is today. This is almost exactly where Mars orbits at present, but by then Mars will have drifted even farther out from the Sun.

While all this is going on, the brightness of the Sun will increase to 2,800 times its present value in the first phase of expansion, and 4,200 times its present value the second time it swells up and becomes a giant star. But even when the star is at its most luminous, the temperature at the surface will have dropped by more than half, from the present value of 5,800 K to only 2,700 K.

None of the new insights offer much hope for the inner planets, Mercury and Venus. Mercury is so close to the Sun that it will be swallowed up long before the star reaches its maximum size, and although the orbit of Venus will expand from its present 108 million kilometer radius to 134 million kilometers by the time the Sun reaches its maximum size in the first phase of expansion, that will still be 30 million kilometers below the surface of the Sun. Drag from the gases in the atmosphere of the Sun will quickly send the planet Venus spiraling inward to a fiery doom.

It's a moot point how long it will take for the Earth to become uncomfortably warm for our kind of life, but we can use these calculations to put one of the problems now facing humankind in a cosmic perspective. It is now well established that the anthropogenic greenhouse effect is likely to raise the average temperature of our planet by a minimum of 5 degrees Celsius by the end of the twenty-first century (indeed, this is a quite conservative estimate). The gradual warming of the Sun as it ages would produce the same warming over an interval of about 800 million years. In other words, in round numbers human activities are speeding up the process by a factor of ten million. The Sussex team suggest that the Earth might be regarded as uninhabitable for life-forms like ourselves by the time the oceans start to boil, which will happen in about 5.7 billion years, assuming we stop interfering with the heat balance of our planet.* That is, perhaps, long enough for our descendants or (more likely) any new intelligent species that develops on Earth to find a new home in space; but there is another partial solution to the problem, which is perhaps not to be taken seriously but which shows how small effects can add up over astronomical timescales.

As anyone who has an interest in the exploration of the Solar System using unmanned space probes knows, these probes are often given a boost on their journeys to distant planets by being sent on a "slingshot" orbit around one of the other planets — perhaps Venus or Jupiter — which uses the gravity of the planet to accelerate the probe on its way. Since you can't get something for nothing in this Universe, this means, of course, that the planet loses a corresponding amount of energy. But since the mass of a space probe is tiny compared with the mass of a planet, the effect is so small that it can be ignored. You could turn the process around and send a space probe on a trajectory past a planet which made the space probe slow down and thus gave a minuscule amount of energy to the planet. But what would be the point?

---

*When this happens, of course, worlds farther out from the Sun, such as Jupiter's icy moon Europa, will become warmer and may provide conditions suitable for life. Some astronomers have therefore suggested that the search for life in space should include investigation of planetary systems around subgiant stars, not just Sun-like stars.

Well, if the "space probe" were big enough and the planet were the Earth, there just might be one.

Just for fun, at the beginning of the twenty-first century a team of American researchers calculated how much effort it would take to change the speed of the Earth in its orbit in such a way that it gradually moved outward to compensate for the gradual warming of the Sun as it ages. They found that the job could be done using existing technology and some very careful long-term planning. The trick involves taking an asteroid with a diameter of about a hundred kilometers (five times bigger than the one thought to have killed off the dinosaurs) and attaching rocket motors to it to steer it into an orbit which sent it whipping past the Earth on just the right trajectory. In a long elliptical orbit stretching out past Jupiter and Saturn, the space rock could be sent zipping past the Earth every six thousand years or so, each time giving our planet a boost and moving it outward by as much as a few kilometers. At the other end of its orbit, the asteroid would gain energy from Jupiter or Saturn, maintaining its own orbit but slightly shrinking theirs. The net effect would be that the Earth took energy from the outer planets and crept steadily outward from the Sun. In principle, this could keep our planet comfortable until the Sun became a red giant.

The scenario is not without its hazards — for a start, the space rock would have to come within 15,000 kilometers of the Earth on every pass, and a tiny error could lead to a catastrophic impact. But when you consider that 2007 marks only the fiftieth anniversary of the launch of the first artificial Earth satellite, Sputnik 1, and we already have the technology to do the trick, maybe if we do survive the next century or so our descendants will be in a position to do a little more sophisticated planetary engineering.

In the very long term, though, the Sun itself will die and end up as a cooling white dwarf. Stars that start out with significantly more mass than our Sun may end up as even more compact neutron stars, where rather more mass than our Sun possesses is packed into the size of a large mountain on Earth, or even as black holes. Everything dies. We know far less about the ultimate fate of the Universe than we do about its beginnings, but we *think* we know what would happen to matter if the Universe carried on expanding forever, and there are some intriguing speculations (still based in real science, for all their speculative nature) about what might happen if it does not. These are inevitably the most speculative topics discussed in this book, but they are no more speculative than ideas about how it all began were a single human lifetime ago.

When we talk about the fate of matter, we can only really discuss the fate of baryonic matter, since we don't know what the rest of the material Universe is. If the expansion of the Universe continues for long enough, eventually star formation will come to an end as all but a trace of the star-making material is used up. The timescales involved are so long that it is hardly worth bothering to spell them out, but this process should have ended within a few trillion (10^{12}) years from now — that is, when the Universe is about a hundred times older than the time that has elapsed so far since the Big Bang. Galaxies will fade away as all their stars become either white dwarfs (cooling into blackness), neutron stars, or black holes. They will also shrink, partly because they will lose energy through gravitational radiation, and partly because of encounters between stars in which energy is exchanged (like the exchange of energy in a gravitational slingshot trajectory) so that one star gains energy and is ejected into intergalactic space while the other loses energy and moves into a tighter orbit around the galactic center. Most galaxies already harbor black holes in their hearts, and this process will cause these black holes to grow and engulf more and more matter.

Even baryons that survive this fate will not last forever. As we discussed earlier, the same processes that allowed baryons to form in the Big Bang require them to be unstable in the very long term. The ultimate baryonic particles are protons and electrons (even neutrons decay into protons, neutrinos, and electrons on a timescale of minutes), and protons must (according to our present understanding of the particle world) themselves decay, essentially into positrons and energetic radiation, on a timescale of 10^{32} years or more. Since there is a balance between all the positive charge in the Universe and all the negative charge in the Universe, by the time this process is complete, perhaps in 10^{34} years from now, all the baryons in the Universe will have been converted into neutrinos, energy, and an equal number of electrons and positrons. The electrons and positrons in their turn will inevitably meet one another and annihilate in a flash of gamma rays. There will still be "matter" around in the form of supermassive black holes, made out of all the rest of the original baryons, but even a black hole isn't forever. Through a process known as Hawking radiation, the energy of a large black hole is very slowly converted into radiant energy plus equal amounts of particles and antiparticles, which will themselves meet one another and be annihilated. (This happens more quickly for smaller black holes, so they will be long gone by the time protons have decayed.) In about 10^{120} years, if the Universe exists that long, everything will have evaporated through the Hawking process.

But will the Universe last that long in anything like its present form? The smart money today says it will not, although nobody is yet smart enough to say which of three possible alternative futures is more likely.

On the old picture of cosmology—and "old" in this case means

more or less pre-2000 — the Universe might have kept expanding forever, but at a slower and slower rate, allowing all these processes of
decay and annihilation to take place. But the presence of dark energy,
or the cosmological constant, changes all that. One other feature of
the old cosmology is that there would be ample opportunity for any
intelligent species that existed to observe the fate of the Universe. The
volume of the Universe that we can see is, in a real sense, expanding at
the speed of light. The distance we can see out into the Universe is the
distance that light has been able to travel in the time since the Big
Bang, and this volume of space is getting bigger at the speed of light.
Beyond this limit, there may be regions of the Universe that are
receding from us faster than the speed of light (because space itself is
expanding, not because they are moving through space faster than
light), of which we can have no knowledge. With the Universe expanding ever more slowly — but with the "light bubble" always moving outward at the speed of light, even though clusters of galaxies
might end up separated by vast distances — on the old picture it was
possible to imagine super-sensitive detectors watching the events in
those distant galaxies right up until matter itself decayed. But that
picture is no longer viable.

We know (or at least, we *think* we know) that the expansion of the
Universe is accelerating today because of the presence of dark energy.
The simplest interpretation of the evidence is that this is associated
with a cosmological constant that really is constant — that each volume of space has an inherent, fixed amount of dark energy associated
with it. This interpretation of the acceleration is supported by the
latest observations (at the time of writing) of distant supernovas,
from a project called the Supernova Legacy Survey. In December
2005, first results from this survey came out exactly in line with the
idea that the acceleration of the expanding Universe is indeed driven

by a cosmological constant, or Lambda force. If that is the case, the processes of decay and annihilation will proceed in much the same way that we have already described, but there will not be much chance for intelligent observers to study them. Because the expansion of the Universe is steadily accelerating, it will not only carry distant clusters of galaxies out past the surface of the light bubble (sometimes referred to as the cosmic horizon) faster than the light bubble itself can expand, but it will do so faster and faster as time passes. It is very simple to calculate that if the acceleration of the Universe continues at the same rate we can see today, then every galaxy beyond the small Local Group of which our Milky Way is a member will have been carried out of sight within a couple of hundred billion years, a little more than ten times the present age of the Universe. What remains of our Galaxy will still be surrounded by a bubble of visible space in which the cosmic horizon is receding at the speed of light, but there will be nothing left to see in all the "visible" Universe.

But what if the cosmological constant isn't really constant after all? What if the amount of dark energy associated with a particular volume of space changes as time passes, either getting bigger or getting smaller? Our observations of the way distant galaxies are receding from us, revealed by the supernova studies described earlier and studies of the cosmic microwave background radiation, set fairly strict limits on how rapidly these parameters could be changing, but these limits leave room for some intriguing speculations.

The first possibility is that the strength of the dark energy is increasing as time passes. There is a certain logic behind this speculation, since it would explain why the cosmological constant is so small today—if it started out as zero and has slowly gotten bigger, it can't help but pass through a stage where it is very small. But it doesn't stop there. The idea presents a dramatic vision of the future which is an extreme variation on the scenario we have described so far. In its most extreme form, this suggests that instead of being at an early stage in the life of the Universe, we may already be almost halfway from the Big Bang to the time when the Universe as we know it ends, but that intelligent observers will have ample opportunity to watch how it all happens. This scenario goes by the dramatic name of the Big Rip, and its proponents like to increase the drama by referring to the extra dark energy that makes the expansion run away exponentially fast as "phantom energy"; actually it is the same dark energy we discussed before, just more of it.*

On this picture, the expansion of the Universe feeds back on itself, with the expansion triggering the growth of dark energy and the increasing strength of dark energy making the expansion accelerate even more rapidly. On the conventional picture, the cosmological constant stays small, so that in systems like the Sun, stars, and Milky Way held together by gravity, there is no expansion — gravity overwhelms dark energy. But according to the Big Rip idea, eventually the expansion of the Universe will dominate, first over gravity, then over the other forces of nature, even on the smallest scales. The most extreme version of this scenario allowed by the observational constraints suggests that the end could come roughly 21 billion years from now. But because the runaway expansion grows exponentially, nothing too dramatic will happen for most of that time, with all the violent activity confined to the last billion years or so of the life of the Universe.

Because of the acceleration of the acceleration of the Universe, the

---

*Phantom energy would also be useful in another context. The "wormholes" that are used in science-fiction stories such as "Stargate" as shortcuts through space and time are unlikely to exist in the real Universe because gravity would snap them shut. But if phantom energy exists, it might be used to hold wormholes open against the pull of gravity.

time when all the galaxies of the Local Group are freed from their gravitational bonds to one another is brought forward to about 20 billion years from now, just one-tenth of the time it takes to get to that stage with uniform acceleration at the present rate. That is soon enough that our Galaxy could still exist as a recognizable island in space, although it is likely to have been considerably enlarged and changed by a merger with the nearby Andromeda galaxy (also known as M31). Our Solar System will long since have gone, of course, but it is reasonable to imagine that something like our Solar System might exist in the aging supergalaxy at that time, with intelligent beings on an Earth-like planet orbiting a Sun-like star. When their home Galaxy began to fly apart as the repulsion of dark energy overcame the gravitational attraction between the stars, they would be just 60 million years away from doomsday, roughly the same time that has elapsed from the death of the dinosaurs to the present day — but at that time the distance to the cosmic horizon would still be about 70 megaparsecs (some 230 million light years), so there would be time after the home Galaxy was "stripped" for light to arrive from other nearby galaxies (the remnants of the Local Group) and for observers to study the way in which those galaxies had blown apart.

Three months before the end, the planets of that future "solar system" would fly free from their parent star, and any survivors of that catastrophe would find their home planet itself exploding into its component atoms about thirty minutes from the moment of doom. Atoms would be torn to pieces in the final 10^{-19} seconds, leaving an expanding, flat, featureless void. Perhaps that's all there is — the end of time. Perhaps these are just the conditions that might trigger a new phase of inflation. Just possibly, this vision of the far future is also a vision of how the Universe as we know it began. Pure speculation; but intriguingly the second alternative fate for the Universe suggests an even closer link between the beginning of time and the end of time. It goes by the name of Big Crunch, and it stems from the possibility that the dark energy may get weaker, not stronger, as time passes.

The first variation on this theme looked very much like the old twentieth-century scenario of an ever-expanding universe in which galaxies gradually faded away and matter decayed. If the cosmological constant gradually declined to zero, we would end up in pretty much the same situation as if it had always been zero from the start. But why should the decline stop at zero? As far as the cosmological equations are concerned, if the strength of the dark energy can decrease to zero from a positive value, then it can carry on decreasing to take negative values. And just as a *positive* dark energy opposes gravity and makes the expansion of the Universe accelerate, so a *negative* dark energy adds to the power of gravity, first slowing the expansion of the Universe and then making the Universe contract at an accelerating rate. If that is the case, then with the most extreme rate of decrease allowed by the observations so far, we could be almost exactly halfway through the life of the Universe, with the ultimate collapse back into a singularity due to take place 12 to 14 billion years from now (equally, though, cosmic doomsday may be postponed until 40 billion years from now).

Nothing very dramatic would be visible to any intelligent observers around in our Galaxy at the time the expansion of the Universe turned around and became a collapse, but the change would happen everywhere in the Universe at once. Because of the finite speed of light, though, shortly after the turnaround observers would see nearby galaxies blue shifted while distant galaxies were still red shifted; the "blue-shift horizon" would then spread out through the Universe at the speed of light. It is impossible to say exactly when the turnaround would occur, since we don't have enough information about the way dark energy is changing (if it is changing at all), and in any case things would only start to get interesting as the Universe approached the Big Crunch; so once again it makes sense to chronicle the key developments in terms of the time left before doomsday, which can be conveniently expressed in terms of the size of the Universe. In this case, though, those intelligent observers would not survive to see the interesting things that happened in the last few minutes.

Of course, we can't see or measure the whole Universe, which may well be infinite. But since everything changes together, relative changes in size are the same for any chosen region of space; we can think in terms of the size of the visible Universe today, and consider how this volume of space would contract and what would happen inside it as it did. For a long time, observers would be able to watch the Universe collapse without their own environment being significantly affected. They would see clusters of galaxies falling toward one another and merging, and even mergers between individual galaxies, without life on some hypothetical future Earth becoming uncomfortable. The threat to life on such a planet would come not from these violent and spectacular interactions but from the slow, insidious rise in temperature of the background radiation, which would be blue shifted to higher and higher energies as the Universe contracted. There is so much space between the stars in a galaxy that even when galaxies started to merge, collisions between individual stars would be rare. At the time galaxies began to merge with one another, the temperature of the background radiation — the temperature of the sky — would be no more than about 100 K. And at that time, the Universe would be about one-hundredth of its size today.

From then until the time when the Universe, halving in size every few million years, had shrunk to one-thousandth of its present size, planetary life would become first uncomfortable, then difficult, then impossible. The temperature of the sky would soon reach 300 K, greater than the melting point of ice, and any planetary ice sheets or glaciers would melt. As the temperature of the universal radiation (it hardly seems appropriate to call it "background" radiation anymore) continued to rise, the whole sky would begin to glow, first a dull red and then, at a few thousand Kelvin—comparable to the temperature at the surface of the Sun—orange. With the seas long since boiled away and the atmosphere becoming disrupted as atoms were broken apart into a plasma of nuclei and electrons (decombination), life as we know it would have become impossible on such a planet by the time the Universe was one-thousandth of its present size. It's no coincidence that this decombination would occur when the Universe was the same size that it was when recombination occurred in the expanding Universe; this scenario is simply the Big Bang in reverse.

The closer we got to the Big Crunch, the faster the Universe would collapse and heat up, as we began to see the fireball stage of the Big Bang in reverse. At one-millionth of its present size the Universe would be so hot that stars would explode as the temperature reached several million degrees, comparable to the temperatures inside stars today. At a billionth of the present size, the temperature would reach a billion degrees, and complex nuclei such as oxygen and iron, painstakingly built up inside stars over billions of years, would be blown apart into their constituent protons and neutrons. At a trillionth of the present size, protons and neutrons themselves would disintegrate, at a temperature of about a trillion (10^{12}) degrees, into a quark soup. We would now be only seconds away from the ultimate collapse into a singularity—or from the time when the laws of physics that we know broke down and something wonderful happened. But nobody would be around to see it.

So will it be a Big Crunch or a Big Rip? It may be possible to distinguish between the two extreme possibilities, or at least to set tighter limits on what might happen, sometime in the second decade of the twenty-first century. An instrument called the Large Synoptic Survey Telescope, which might (optimistically) begin operating in 2012, is designed to make very accurate, detailed measurements of the way galaxies clump together, providing sufficient statistical information to set tight constraints on the strength of dark energy, and perhaps on how this has changed as the Universe has aged. And a satellite known as SNAP (for Supernova Acceleration Probe), even more optimistically scheduled to fly before 2015, will provide information about thousands of supernovas in far-distant galaxies, yielding a much more precise measurement of the way the universal expansion is accelerating than we can make today.

Meanwhile, we have only our imaginations to guide us. Fortunately physicists have fertile imaginations, and I have saved my favorite among the current crop of speculations about how it will all end for last. This is the full version of the ekpyrotic universe model mentioned earlier. It introduces dark energy in a natural way; includes elements of both the Big Rip and the Big Crunch, set in the context of a literally eternal cycle of birth, death, and rebirth; and is based on the latest ideas involving M-theory and branes. This doesn't necessarily mean that it is right. But it is a beautifully put together package, and it shows more clearly than any other example just where physics is going in the first decade of the new millennium.

The one ugly thing about the model is its name. It comes from the Greek word for conflagration (*ekpyrosis,* as in pyrotechnics or pyromania), which is appropriate in some ways, as we shall see; but somehow it doesn't trip off the tongue as easily as Big Bang and all the other "Bigs." "Big Splat" has been suggested by some who are critical of the model, slightly to the irritation of its proponents, who remember how the term Big Bang was coined by a critic of that model (Fred

Hoyle) and then stuck. But at least Big Splat does give you a mental picture of what is going on — the collision between two branes, described by the equations of M-theory. Strictly speaking, the ekpyrotic universe is the term used to describe a model of a single collision between branes, producing the Big Bang; but it has now been extended to encompass models in which such collisions recur repeatedly in an endless cycle of death and rebirth. So perhaps "Phoenix model" would be a better term.

Each of the branes involved in this cycle can be thought of in spatial terms as a complete, infinite, three-dimensional universe like our own, with time as the fourth dimension. These universes are separated from one another in a fourth spatial dimension (the fifth dimension altogether). As usual, it helps to think of the equivalent two-dimensional brane and its neighbor, from which it is separated in the third spatial dimension, like two adjacent pages in a book. As we explained earlier, all the familiar particles and forces except gravity are constrained to move and act within a single braneworld (our Universe), but gravity can leak through the fifth dimension to influence the universe next door. Because the process is cyclic (and possibly endless), we can start describing it at any point in the cycle. The logical place seems to be the point in the cycle which corresponds to what we usually think of as the Big Bang — since, as we shall see, this also corresponds to the end of the Universe as we know it.

Imagine two branes approaching each other along the fifth dimension, like two sheets of paper being brought together face to face. From the perspective of our brane, if intelligent beings were around at that time and had instruments that could "look" along the fifth dimension they would simply see the other brane approaching them, as if their home brane (our brane) were standing still. The branes would be attracted to one another along the fifth dimension by forces which act along that dimension — essentially, the real, full-strength force of gravity. For reasons that will shortly become clear, at this stage of the cycle both branes would be essentially empty and extremely flat, in terms of their space-time curvature. But no space-time can be *completely* flat, because of quantum effects; so there would be irregularities in each of the branes, the equivalent of hills and valleys on a two-dimensional surface. Even though the two branes approached each other like parallel sheets of paper being brought into contact, they would touch first at places where there were bumps in the membranes. You might think that these irregularities would be tiny, and they do start out that way; but according to the equations of M-theory, as the two branes came very close together powerful forces would act both to pull them strongly into contact and to enlarge the quantum ripples.

This would produce irregularities of all sizes — such irregularities are said to be "scale invariant." Some would still be subatomic in size, others might be as big as the entire visible Universe today, with all the in-between possibilities as well. But the ones that are of interest to us would have been about a meter across at the time the two membranes came into contact, because the calculations show that these would have been just the right size to have produced a universe like the one we see around us.

Because the membranes are spatially three-dimensional, the best way to picture this event is to imagine a spherical volume of space, with a diameter of a meter or so, within which every point in one three-dimensional volume (every point in 3-space) instantaneously comes into contact with every point in the 3-space of the other brane. The result would be a fireball of energy that expanded dramatically in the three spatial dimensions of "our" brane, growing exponentially but at a rate which slowed down as time passed. Crucially, this would *not* involve a period of inflation, although initially the expansion would be dramatic enough by present-day standards. At first, the universe would double in size once every 10^{-20} seconds, but the rate of expansion would slow down so that today the doubling time would be about 10^{10} years; in the future the doubling time would decrease again as the dark energy made the expansion of the universe accelerate. Overall, on this scenario the universe would grow by a factor of more than 10^{20} (a hundred billion billion) or even 10^{30} as it is "born," which explains the observed extreme flatness of our Universe in the same way that inflation does.

On this picture, our entire Universe has expanded from this meter-wide fireball. But during the process of the collision itself, a second set of quantum ripples was imprinted on space-time, producing the ripples in the cosmic microwave background radiation and providing the seeds on which clusters of galaxies could grow, just as in the standard picture of inflation. Crucially, though, there is no singularity at the beginning of time, and "our Universe" was never less than a meter across, and never experienced infinite density; it does "begin" at a temperature of 10^{24} degrees, but, large though that number is, it is still finite. This is one of the main reasons why many cosmologists find the model attractive. It also means that "our Universe" is not unique, even if we think only in terms of our brane. There must be many other regions of the brane (infinitely many, if the branes are infinite) which came into contact and have given birth to expanding universes in this way. But they will be forever beyond the horizon of our visible bubble of expanding space, because the fabric of the membrane between the universes also expands as time passes. It does so because of the way the two membranes recoil from each other after they collide.

The other membrane suffers a slightly different fate after the two branes rebound from each other and start moving apart along the fifth dimension. According to this model, the extra dimension is warped, in such a way that as you travel along the extra dimension in one direction the size of the familiar three spatial dimensions increases, while if you move in the other direction the size of the familiar three dimensions decreases. When the branes rebound from the collision, "our" Universe happens to move in the direction of expanding 3-space. Initially, the universe next door rebounds the other way along the fifth dimension and contracts. But the force of attraction between the branes (the real force of gravity) is so powerful that the second brane gets turned around and dragged along behind our brane, so that it soon starts expanding as well. The two branes each go through the process of expansion, dilution, and matter decay that we described for an eternally expanding universe, but they have no contact with each other for all that time — a time measured in trillions (multiples of 10^{12}) years. The separation between them is tiny by any human standards — perhaps a few thousand Planck lengths, or 10^{-30} centimeters (far smaller than the diameter of a proton); but as the proverb says, a miss is as good as a mile. Thanks to the interbrane force, however, as well as becoming empty and flat the two branes are kept parallel to each other the whole time.

For most of that time, the two branes "hover" at about the same distance apart. Gradually, however, the interbrane force of attraction overcomes the kinetic energy imparted by the bounce, and pulls them back together. This process starts very, very slowly indeed, but speeds up dramatically as the branes get closer and closer together. Unlike the force pulling a stretched spring back to its normal size, the force of attraction between the two branes is much stronger when the branes are closer together than it is when they are a few thousand Planck lengths apart. The branes themselves (by now consisting of essentially empty, flat space-time) contract slightly during this phase of the cycle, but only by a factor of 10 or so (compared with the factor of up to 10^{30} they stretched during the expansion phase), and quantum ripples in their space-times are enlarged up to all scales, including meter-sized bumps.

The energy for the Big Bang comes directly from the kinetic energy of motion of the two branes when they collide,* like a clash of cymbals, and because of the way gravity is spread out into the fifth dimension, as the space between the membranes shrinks in that direction the effective strength of gravity itself increases in each of the branes — this means that at the "moment of the Big Bang" in our Universe, when the Universe was a meter in diameter, Newton's gravitational constant was larger than it is today. But it had fallen to its present value within a fraction of a second.

We have completed the cycle, and this is where we came in. The model produces exactly the same picture of the Universe we live in as the standard "ΛCDM" model with inflation, including the nature of the background radiation, but without the infinities that are the troubling feature of that model. The dark matter required to explain the behavior of visible matter in our Universe could be weakly interacting particles in our Universe or the influence of particles on the other membrane leaking through into ours; the model does not distinguish between the two possibilities. But the dark energy that is making the expansion of the Universe accelerate today is an essential feature of the model. The average density of matter, the temperature, and all other physical properties of the universe are identical at each bang (each meter-sized fireball) and at the corresponding points in each cycle; the universe just refills itself every time, even though the quan-

---

*Because no energy is lost in the process (there is nowhere for it to go!), the energy of the bounce is the same as the energy of the collision, which is why the cycle can repeat indefinitely. In a sense, there is no friction involved to dissipate energy.

tum fluctuations themselves are statistically different in every cycle, because each bang is associated with a single quantum fluctuation. But the conditions which make the Universe what it is today were imprinted at the end of the collapsing phase of the previous cycle, not at the beginning of the present expansion phase.

There is one subtle, but important, difference between the Phoenix model and inflation. The standard model involving inflation predicts that the Universe should be filled with gravitational radiation that has left an imprint on the cosmic microwave background; the Phoenix model predicts that there will be no such gravitational wave effect observable in the cosmic microwave background. Although space probes capable of measuring the background radiation to the required precision are unlikely to fly much before 2020, this does mean that, like all good scientific ideas, this one can be tested by experiment.

Which of these scenarios for how it will all end you prefer is at present entirely a matter of choice. As of 2005, we have no experiments or observational evidence to distinguish among them. That will surely change in the not too distant future; but my own favorite is the Phoenix model of eternally cycling universes in which death is always followed by rebirth. As it happens, the Phoenix scenario echoes a vision dating back more than two hundred years, a word-picture painted by Erasmus Darwin, the grandfather of Charles Darwin, in his portrayal of scientific speculations in *The Botanic Garden,* first published in 1791:

Roll on, ye Stars! exult in youthful prime,
Mark with bright curves the pointless steps of Time;
Near and more near your beamy cars approach,
And lessening orbs on lessening orbs encroach; —
Flowers of the sky! ye too to age must yield,

Frail as your silken sisters of the field!
Star after star from Heaven's high arch shall rush,
Suns sink on suns, and systems systems crush,
Headlong, extinct, to one dark center fall,
And Death and Night and Chaos mingle all!
— Till o'er the wreck, emerging from the storm,
Immortal Nature lifts her changeful form,
Mounts from her funeral pyre on wings of flame,
And soars and shines, another and the same.

To be sure, the image is more like the traditional Big Crunch than the modern cyclic universe; but it's not bad for an eighteenth century speculation. If nothing else, it does at least enable me to end my book where the story started — in the beginning.

Glossary

Antimatter: Mirror-image matter in which properties such as electric charge are the opposite of those for everyday particles. So, for example, the antimatter counterpart to the electron has positive charge rather than negative charge.

Astronomical unit: Unit of distance used by astronomers, equivalent to the average distance from the Earth to the Sun.

Axion: Hypothetical subatomic particle which may contribute a large proportion of the overall mass of the Universe.

Baryon: The kind of entity that we are used to thinking of as "particles," things like protons and neutrons, but excluding electrons.

Beta decay: The process in which a neutron ejects an electron and is converted into a proton.

Black hole: a concentration of matter in which the gravitational field is strong enough to bend space-time around upon itself so that nothing, not even light, can escape.

Blueshift: Squashing of the wavelength of light in the spectrum of an object moving toward the observer.

Bosons: The family of particles associated with what we usually think of as forces, such as electromagnetism. The W^+, W^-, and Z bosons are the particles associated with the weak nuclear force.

Bottom: A name given to a property of quarks. The opposite of Top.

Charm: A name given to a property of quarks. The opposite of Strange.

Classical physics: The physics that applies on large scales, more or less scales larger than the size of atoms.

Cold Dark Matter (CDM): Matter known to contribute a significant amount of the mass of the Universe, but not in the form of baryons and leptons. Nobody yet knows exactly what CDM is.

Critical density: The density for which the space-time of the Universe is flat.

Dark energy: A form of energy which fills the Universe and is chiefly responsible for making space-time flat. Nobody yet knows exactly what the dark energy is; it may also be responsible for making the expansion of the Universe accelerate.

Doppler effect: Overall name for blueshift and redshift.

Down: A name given to a property of quarks. The opposite of Up.

Electron: Light, negatively charged particle; on Earth, most electrons are bound up in atoms along with protons and neutrons.

Electron volt: Unit used by particle physicists to measure mass and energy. The mass of a proton is roughly 1 Giga electron Volt, or 1 GeV.

eV: Electron volts.

Fermion: The kind of particle everyday matter is composed of. Baryons and leptons are fermions.

Field: The region of influence of a force, such as gravity or electromagnetism.

Galaxy: With a small "g," any one of hundreds of billions of islands of stars in the Universe. With a capital "G," our home in space, the Milky Way Galaxy, containing several hundred billion stars.

GeV: Gigavolt (1 billion electron volts); 1 GeV is roughly the mass of a proton or a hydrogen atom.

Grand Unified Theory (GUT): Any theory (I prefer the term *model*) which attempts to unify the description of all the forces of nature except gravity in one mathematical package.

Gravitino: Supersymmetric counterpart to the graviton.

Graviton: Particle associated with gravity. Gravitons are members of the boson family.

Higgs field: Hypothetical field, thought to fill the entire Universe, which gives mass to particles.

Inflation: Popular model of the very early Universe which explains many observed features as a result of extremely rapid (exponential) expansion of space-time in the first fraction of a second of time.

Interaction: Term used by physicists to refer to the forces of nature.

Kaons (K-particles): Family of three bosons whose properties demonstrate that there are small asymmetries in the laws of physics. These asymmetries allow matter to exist.

keV: One thousand eV.

K meson: Kaon.

Lepton: A family of particles that includes the electron and the neutrinos.

Light year: The distance traveled by light in one year — a measure of *distance* not of time.

Lookback time: The time taken for light to reach us from a distant object. If a galaxy is ten million light years away, we see it by light which started on its journey ten million years ago. The farther we look out into the Universe, the farther back in time we see.

Meson: A subgroup of bosons.

MeV: One million eV.

Muon: A heavier counterpart to the electron.

Muon neutrino: A heavier (but still very light) counterpart to the neutrino.

Neutralino: A little neutral SUSY particle; a generic name, rather than a specific particle.

Neutrino: Very light lepton with no electric charge.

Neutron: Relatively massive, electrically neutral particle; on Earth, most neutrons are bound up in atoms along with protons and electrons.

Nucleon: Generic name for protons and neutrons.

Nucleus: Inner core of an atom, composed of protons and neutrons (together known as nucleons).

Oscillation: The way in which members of some families of particles (for example, neutrinos) can change from one form to another and back again.

Photino: SUSY counterpart to the photon.

Photon: Particle (boson) associated with electromagnetism.

Pion: Family of three bosons involved in interactions between protons and neutrons.

Positron: Antimatter counterpart to the electron.

Proton: Relatively massive, positively charged particle; on Earth, most protons are bound up in atoms along with electrons and neutrons.

Quantum (adj.): Referring to the world of the very small, where the rules of quantum physics apply.

Quantum (noun): The smallest amount of something that can exist. For example, the quantum of the electromagnetic field is the photon.

Quantum field theory (= quantum theory of forces): Any theory which describes the interaction between matter particles (fermions) in terms of the exchange of field quanta (bosons). In some, but not all, cases (notably gravity) the field quanta can also interact among themselves.

Quantum physics: The physics that operates on small scales, more or less the scale of atoms and below.

Quark: Family of fundamental particles out of which all baryons are made.

Quasar: The active core of a galaxy, probably fuelled by matter being swallowed by a black hole with hundreds of millions solar masses. Most quasars outshine the entire galaxy that surrounds them, literally brighter than a hundred billion Suns. This makes them visible far away across the Universe. The name is a contraction of "quasi-stellar" object because they look like stars in photographs of distant regions of the Universe.

Redshift: Stretching of the wavelength of light in the spectrum of an object moving away from the observer.

Selectron: SUSY counterpart to the electron.

Solar System: The Sun and its family of planets, comets, and other cosmic debris.

Strange: A name given to a property of quarks. The opposite of Charm.

Supersymmetric partner: The counterparts to everyday bosons and fermions predicted by the idea of supersymmetry (SUSY).

Supersymmetry: A model in which every variety of fermion has a bosonic counterpart, and every variety of bosom has a fermionic counterpart. This is part of an attempt to find a Theory of Everything (TOE).

SUSY: Supersymmetry.

Tau: Heavier counterpart to the electron.

Tau neutrino: Heavier (but still very light) counterpart to the neutrino.

TeV: Thousand billion eV.

Theory of Everything (TOE): Any theory (I prefer the term *model*) which attempts to unify the description of gravity and all the other forces of nature in one mathematical package.

Top: A name given to a property of quarks. The opposite of Bottom.

Up: A name given to a property of quarks. The opposite of Down.

White Dwarf: A star at the end of its life that has settled into a solid ball about the size of the Earth, containing roughly as much mass as our Sun.

WIMP: Acronym for Weakly Interacting Massive Particle, another term for Cold Dark Matter.

Bibliography

The appearance of several of my own books in this list is not intended to indicate that they are the best books ever written on the subject but that they complement the present book by providing background on topics briefly touched on here.

Abbott, Edwin. *Flatland*. 1884. London: Shambhala, 1999.

Allday, Jonathan. *Quarks, Leptons and the Big Bang*. Bristol, Eng.: IoP Publishing, 1998.

Arnett, David, "Sir Fred Hoyle and the Theory of the Synthesis of the Elements." In *The Scientific Legacy of Fred Hoyle*. Ed. D. Gough. Cambridge: Cambridge University Press, 2005.

Barrow, John, and Frank Tipler. *The Anthropic Cosmological Principle*. Oxford: Oxford University Press, 1986.

Bartusiak, Marcia. *Einstein's Unfinished Symphony*. Washington, D.C.: Joseph Henry Press, 2000.

Dyson, Freeman. *Origins of Life*, 2d ed. Cambridge: Cambridge University Press, 1999.

Feynman, Richard. *The Character of Physical Law*. Cambridge: MIT Press, 1965.

Greene, Brian. *The Elegant Universe*. London: Jonathan Cape, 1999.

———. *The Fabric of the Cosmos*. London: Allen Lane, 2004.

Gribbin, John. *The Birth of Time*. London: Weidenfeld and Nicolson, 1999.

———. *Deep Simplicity*. London: Allen Lane, 2004.

———. *In the Beginning*. London: Viking, 1993.

———. *Stardust*. London: Allen Lane, 2000.

Gribbin, John, and Mary Gribbin. *Annus Mirabilis*. New York: Chamberlain, 2005.

Gribbin, John, and Martin Rees. *The Stuff of the Universe*. London: Penguin, 1993.

Guth, Alan. *The Inflationary Universe*. London: Jonathan Cape, 1997.

Jeans, James. *Astronomy and Cosmogony*. Cambridge: Cambridge University Press, 1929.

Krauss, Lawrence. *The Fifth Essence*. New York: Basic, 1989.

Mitton, Simon. *Fred Hoyle*. London: Aurum, 2005.

National Research Council. *Connecting Quarks with the Cosmos*. Washington, D.C.: National Academies Press, 2003.

Oparin, A. I. *The Origin of Life on the Earth*. 3d ed. Edinburgh: Oliver and Boyd, 1957.

Rees, Martin. *Just Six Numbers*. London: Weidenfeld and Nicolson, 1999.

Schröder, Peter, Robert Smith, and Kevin Apps, "Solar Evolution and the Distant Future of the Earth," *Astronomy and Geophysics* 42 (December 2001): 26–29.

Smolin, Lee. *The Life of the Cosmos*. London: Weidenfeld and Nicolson, 1997.

Stewart, Ian. *Flatterland*. London: Pan, 2003.

Thorne, Kip. *Black Holes and Time Warps*. New York: Norton, 1994.

Wickramasinghe, Chandra, Geoffrey Burbidge, and Jayant Narlikar. *Fred Hoyle's Universe*. Dordrecht: Kluwer, 2003.

Index